Weather and Water

Developed at
The Lawrence Hall of Science,
University of California, Berkeley
Published and distributed by
Delta Education,
a member of the School Specialty Family

© 2018 by The Regents of the University of California. All rights reserved. No part of this book may be reproduced or transmitted in any form or by any means, electronic or mechanical, including photocopying or recording, or by any information storage and retrieval system, without prior written permission.

1558515
978-1-62571-786-3
Printing 4 — 2/2019
Webcrafters, Madison, WI

Table of Contents

Readings

Investigation 1: What Is Weather?
Severe Weather....................3
What's in the Air?................18
A Thin Blue Veil..................24

Investigation 2: Air Pressure and Wind
What Is Air Pressure?.............32

Investigation 3: Convection
Density...........................41
Density with Dey..................47
Convection........................51

Investigation 4: Radiation
Seasons...........................53
Thermometer: A Device to Measure Temperature......................59

Investigation 5: Conduction
Home Insulation...................64

Investigation 6: Air Flow
Heating the Atmosphere............69
Wind on Earth.....................76

Investigation 7: Water in the Air
Weather Balloons and the Radiosonde...85
Animal Rains......................87

Investigation 8: The Water Planet
Earth: The Water Planet...........91
Ocean Currents and Gyres..........96
El Niño..........................103

Investigation 9: Climate over Time
Climates: Past, Present, and Future.....105

Images and Data..................111

References
Science Practices.................134
Engineering Practices.............135
Engineering Design Process........136
Science Safety Rules..............137
Glossary..........................138
Index.............................142

Fog is a common weather feature around San Francisco's Golden Gate Bridge. In fact, the bright orange color was chosen because of its visibility in fog, and the bridge is equipped with foghorns.

Severe Weather

"Severe Weather Alert" flashes across the TV screen. Thunderstorms, tornadoes, blizzards, or hurricanes may be on the way. Are you in the path of the storm? Time to prepare for the unexpected.

Weather is fairly predictable most of the time. During the summer months in the San Francisco Bay area, you can expect fog in the mornings and late afternoons, with the possibility of sunshine midday. In the southeastern United States, summer days are often hot and humid. In the Midwest and East, winters are usually cold, cloudy, and snowy. These are the normal conditions that people come to expect where they live.

It is the change from normal that catches people's attention, whether they see it on TV or experience it for themselves. **Tornadoes**, **thunderstorms**, windstorms, **hurricanes**, **typhoons**, and **floods** are examples of what is known as **severe weather**. Severe weather is out of the ordinary. It usually causes dangerous conditions that can damage property and threaten lives.

The catastrophic Johnstown, Pennsylvania, flood was caused by an epic rainstorm and a poorly maintained dam. The entire contents of a human-made lake swept through the unsuspecting town.

Flood

Floods occur when water overflows the natural or artificial banks of a stream or other body of water. The water moves over normally dry land or accumulates in low-lying areas. Floods often happen with heavy rainfall over short periods of time or when a large amount of snow melts quickly.

Floods may also occur behind ice dams on rivers, during very high tides, or after tsunamis (soo•NAH•mees), huge waves caused by earthquakes under water. **Flash floods** are short, rapid, unexpected flows of muddy water rushing down a canyon. They are often caused by thunderstorms over mountains, during which large amounts of water flow down a single canyon.

Johnstown, Pennsylvania, 1889. A flood of disastrous proportions hit the city of Johnstown, Pennsylvania, in 1889. An earthen dam collapsed after heavy rains. A great wall of water rushed down the Conemaugh River valley at speeds of up to 64 kilometers (km) an hour. The wall of water, 10 meters (m) high, devastated the town. It washed away most of the northern half of the city, killing 2,209 people and destroying 1,600 houses.

Drought

Droughts are less-than-normal precipitation over a long period of time. They usually cause water shortages, including low flow or no flow in streams. Soil moisture and **groundwater** levels decrease. Droughts are especially disastrous for agriculture.

The Dust Bowl, United States, 1930s.

An area of the United States that includes parts of Colorado, Kansas, New Mexico, and the panhandles of Texas and Oklahoma was named the Dust Bowl in the early 1930s. A severe drought occurred after years of poor land management. The native grasses had been removed, exposing the topsoil. Strong spring winds blew away the topsoil, causing "black blizzards." Thousands of families left the area at the height of the Great Depression (1929–1939) because of the poor living conditions brought on by the drought.

Pakistan, 2000.
Large parts of Pakistan were hit by drought in 2000. No rain fell in the Balochistan area for more than 3 years. Many people were forced to leave some of the remote villages. They migrated to cities in search of food and water. Distribution of supplies required a major support effort by the Pakistan government and several international relief agencies.

Droughts have serious economic and environmental effects. Hunger, famine, thirst, disease, wildfires, and loss of habitat are some of the results of long-term drought.

Hail

Hail is precipitation in the form of balls of ice. The diameter of hailstones ranges from 0.5 to 10 centimeters (cm). Hail usually forms during thunderstorms when strong **updrafts** (vertical winds) move through **cumulonimbus** clouds in which temperatures are near or below freezing.

Nodaway County, Missouri, 1898. A huge hailstorm struck Nodaway County in northwest Missouri on September 5, 1898. The hail remained on the ground for 52 days and left the fields unworkable for 2 weeks. On October 27, there was still enough hail left in ravines for the local residents to make ice cream.

Selden, Kansas, 1959. On June 3, 1959, a severe hailstorm struck the town of Selden in northwestern Kansas. Hail fell for 85 minutes. The storm covered an area 10 km by 14.4 km with hailstones to a depth of 46 cm. The storm caused $500,000 worth of damage, which was a lot of money in 1959.

Hail can be very large, causing damage to cars, buildings, crops, and livestock when it falls.

Thunderstorms often result in heavy rain, but they may also produce other weather threats like flooding and dangerous lightning. So never ignore a severe thunderstorm warning!

Thunderstorm

Thunderstorms produce rapidly rising air currents, usually resulting in heavy rain or hail along with **thunder** and **lightning**. A thunderstorm is classified as severe when it produces one or more of these conditions.

- Hail 1.9 cm or more in diameter
- Winds gusting in excess of 96 km per hour
- A tornado

West Central Texas, 1996. On October 22, 1996, surface temperatures dropped from 10 degrees Celsius (°C) to around 0°C in an area of west central Texas. Strong rising winds developed. The storm increased in intensity very quickly. Dime- to egg-sized hail fell, as well as large amounts of sleet and light snow.

Fort Worth, Texas, 1995. An isolated severe storm became the costliest thunderstorm in US history when it devastated the area in and around Fort Worth, Texas, on May 5, 1995. More than 100 people were injured, mostly by softball-sized hail that pelted people attending an outdoor festival. Winds reached 100 to 117 km per hour, dropping hail the size of grapefruit in some areas. The large hail and high winds damaged hundreds of houses, businesses, and vehicles. Damage totals reached more than $2 billion.

Investigation 1: *What Is Weather?*

Lightning and Thunder

Lightning is a visible electric discharge produced by thunderstorms. Not everyone agrees *why* lightning happens, but *what* happens is pretty well understood.

Lightning travels from a cloud to the ground. The electric charge moves downward in steps of approximately 46 m called **step leaders**. This flow continues until the charge reaches something on the ground that is a good conductor. The circuit is completed. The whole event usually takes less than half a second, and a lightning bolt can reach 200 million volts.

Since 1989, US meteorologists have used a network of antennas to detect lightning. An average of 20 million cloud-to-ground strikes happen every year over the continental United States.

Thunder is the explosive sound that usually accompanies lightning. The sound is caused by the rapidly expanding gases in the **atmosphere** along the path of the lightning. How loud and what type of sound you hear depends on atmospheric conditions and how far away you are from the flash.

Lightning can be destructive and deadly—and hot, four times as hot as the surface of the Sun. The most common damage is to trees and tall structures: 10,000 wildfires and 6,000 building fires a year.

Durham, North Carolina, 1995. Lightning during an early-evening thunderstorm killed a golfer in Durham, North Carolina, on July 7, 1995. The man took cover from the storm in a wooden shed at a golf course. Witnesses say the lightning first struck a tree and then bounced to the shed. Other golfers sustained scrapes and burns from the accident. Later they said that the lightning was the brightest they had ever seen and that the thunder came immediately.

Villa Gesell, Argentina, 2014. Three people were killed and 22 others were injured when lightning struck a beach during a sudden thunderstorm. Some people took shelter in a tent, but the structure did not protect them from injury.

Did You Know?

Scientists estimate there are 44 lightning strikes on Earth every second of every day.

Investigation 1: What Is Weather?

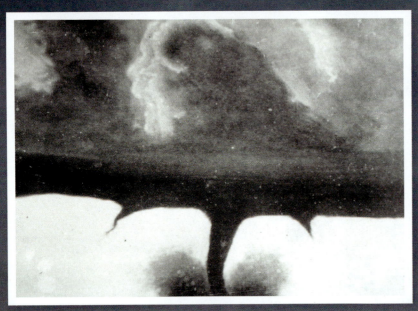

This is the oldest known photo of a tornado. It was taken 35 km southwest of Howard, South Dakota, on August 28, 1884.

Tornado

Tornadoes are rapidly rotating columns of **air** that extend from a thunderstorm to the ground. Wind speeds in a tornado can reach 417 km per hour or more. The path of a tornado on the ground can be more than 1 km wide and 83 km long.

Xenia, Ohio, 1974. On April 3 and 4, 1974, a super tornado outbreak struck the states of Alabama, Georgia, Illinois, Indiana, Kentucky, Michigan, Mississippi, North Carolina, Ohio, South Carolina, Tennessee, Virginia, and West Virginia. Especially destructive tornadoes struck Ohio during the afternoon and early evening of April 3. The town of Xenia was hit the hardest. Thirty people died, and more than 1,100 were injured. More than 1,000 houses were destroyed. The path of damage ranged from 0.4 to 0.83 km wide.

A tornado gets its dark color from the dust picked up by the swirling wind. It gets its name from the Spanish word for thunderstorm, *tronada*.

Oklahoma City, Oklahoma, 1999. Eight thunderstorms produced at least 59 tornadoes in central Oklahoma on May 3, 1999. Many of these tornadoes were very violent and long-lasting. They made direct hits on several populated areas, including Oklahoma City. At least 40 people died in Oklahoma because of the tornadoes, and 675 were injured. The total damage was about $1.2 billion. The tornadoes also caused extensive damage to the Wichita, Kansas, area.

Enterprise, Alabama, 2007. A tornado outbreak spread across the southern United States for 3 days in March 2007, with a total of 56 tornadoes. Enterprise, Alabama, was hit the hardest. Tragedy struck when the tornado went directly through a high school, causing eight fatalities, collapsing part of the school, flipping cars in the parking lot, and tearing trees from the ground. The tornado was estimated to have been 457 m in diameter and to have traveled for 16 km.

Hurricane

Hurricanes are moving wind systems that rotate around an eye, or center of low **atmospheric pressure**. Hurricanes form over warm tropical seas. Wind speeds are more than 119 km per hour in a hurricane. Depending on the strength and location of the storm, it may be called a hurricane, typhoon, or tropical cyclone. Hurricanes and typhoons produce dangerous winds, heavy rains, and flooding. They can cause great property damage and loss of life in coastal areas.

The 1900 Galveston hurricane is the worst weather-related disaster in US history in terms of loss of life. The Weather Bureau issued warnings, but forecasting was untrusted and most people did not evacuate.

Galveston, Texas, 1900. The hurricane that struck Galveston, Texas, on September 8, 1900, is considered to be one of the worst natural disasters in US history. More than 6,000 men, women, and children lost their lives during this great storm. It is estimated that the winds reached 250 to 333 km per hour. The tidal surge was probably 4.6 to 6.2 m. This tide overwhelmed the highest elevation on Galveston Island, which was 2.7 m. More than 3,600 houses were destroyed, with whole blocks of houses totally wiped out, leaving only a few bricks behind.

New Orleans, Louisiana, 2005. Hurricane Katrina was the costliest hurricane in US history, and was the sixth strongest Atlantic hurricane on record. Damage stretched along the Gulf Coast from Texas to central Florida. The levee system failed in New Orleans, leaving the city vulnerable to the huge storm surge and flooding. More than 1,800 lives were lost, and more than 1 million people were displaced from their homes.

Hurricanes first take form as tropical storms. They gather strength and moisture as they travel thousands of miles over warm ocean waters.

Blizzard

Blizzards are severe storms with low temperatures, strong winds, and large amounts of snow. Blizzards have winds of more than 51 km per hour and enough snow to limit visibility to 150 m or less. A severe blizzard has winds of more than 72 km per hour, near-zero visibility, and temperatures of −12°C or lower.

East Coast, United States, 1899. A cold wave hit the East Coast during February 1–14, 1899, causing a huge blizzard and bitter-cold temperatures from the Rockies to the Atlantic Ocean. Snow fell from Louisiana to Georgia and extended northeast into New England. Up to 61 cm of snow fell across much of the Northeast. Winds gusted to 58 km per hour. Florida experienced its first recorded blizzard. Temperatures fell below freezing. Jacksonville, Florida, measured its greatest snowfall ever at just over 4 cm. Tampa, Florida, received its first measurable snow on record.

Blizzards decrease visibility, making travel difficult and dangerous.

Snow and ice can bring down trees and power lines.

Chicago, Illinois, 1967. Chicago, Illinois, set records on January 26 and 27, 1967, with the worst blizzard ever recorded there. More than 52 cm of snow fell in 29 hours and 8 minutes. Trains froze and could not move. Buses were stranded, and O'Hare Airport was closed for 3 days.

Washington, DC, 1979. Known as the Presidents' Day Storm, this blizzard covered much of the Northeast, closing down Washington, DC. Ice formed on the Potomac River. Some mid-Atlantic areas got more than 44 cm of snow. Observers in Baltimore, Maryland, reported snow falling at a rate of 11 cm per hour.

Super Storm, 1993. Twenty-six states were affected by a low-pressure system during March 12–15, 1993. The system brought heavy snow and strong winds. Birmingham, Alabama, received 28.6 cm of snow, while Chattanooga, Tennessee, measured 44 cm by the time the storm passed. Winds gusted up to 83 km per hour as temperatures fell rapidly. Parts of Long Island, New York, recorded wind gusts of 155 km per hour. Over 50 tornadoes hit Florida. The storm claimed more than 200 lives.

A dust storm forms when winds whip over a hot, dry region. Enormous dust clouds, typically 1–2 km high, advance across the land, making it difficult to see or breathe.

Windstorm

Sometimes strong winds occur that are not directly connected with thunderstorms, tornadoes, or hurricanes. **Straight-line winds** are strong winds that have no rotation. The winds can travel at speeds of more than 167 km per hour. If no one is around to observe what happens, it can be difficult to tell if damage came from a straight-line wind or a tornado.

Microbursts are small, very intense downward winds. They affect areas less than 4 km wide. Microbursts can produce winds of more than 280 km per hour. They typically last less than 10 minutes. They are often associated with thunderstorms. Microbursts often cause significant ground damage and are a threat to aviation.

Dust devils are small rotating winds not associated with a thunderstorm. They become visible when they collect dust or debris. Dust devils form when air is heated by a hot surface during fair, hot weather. You are most likely to see a dust devil in arid or semiarid regions such as Texas or New Mexico.

Dust storms are rare conditions in which strong winds carry dust over a large area. They occur when there are drought conditions. In desert areas, sandstorms occur.

Wildfires are driven by winds. Windstorms can spread fire across large areas, making the fire harder to extinguish. Many regions in the world, such as Australia and the southwestern United States, are susceptible to wildfires, so windstorms carry additional risks in these areas.

China, 1998. Parts of China and Mongolia experienced a major dust storm in April 1998. Twelve people were reported missing after the storm, which brought very strong winds. The storm blanketed ten cities and districts. Power and water supplies were cut. A trace of yellow dust carried by the winds was deposited on nearby deserts. Some of the dust was blown across the Pacific Ocean to the United States.

Mars, 2012. Wind and dust can be severe on Mars. A large Martian dust storm can cover most of the **planet**. Mars Reconnaissance Orbiter photographed a dust devil blowing across Mars's northern plains in February 2012. The dust devil in the photograph has a diameter of 30 m.

> **Take Note**
>
> What types of severe weather occur where you live? Make a list in your notebook.

> **Think Questions**
>
> 1. What weather factors might help you forecast a hurricane?
> 2. What weather factors might help you forecast a blizzard?
> 3. What are some forms of technology mentioned in this article that help people forecast severe weather events?

A dust devil on Mars forms the same way one does on Earth. But the dust devils observed by Mars probes are monstrously large!

Investigation 1: What Is Weather?

What's in the Air?

Air? Can't see it. Can't taste it. Can't smell it. If you pay attention, you might feel it as a gentle breeze brushing across your skin.

A strong gust of wind can nearly knock you over. But most of the time, we cannot see, smell, or feel air. Is it one thing, or a mixture? Is it everywhere, or just in some places?

As we go about our daily lives, we usually travel with our feet on the solid Earth and our heads in the atmosphere. The atmosphere completely surrounds us. It presses firmly on every square centimeter of our top, front, back, and sides. We could try to get out of the atmosphere by locking ourselves inside a car or hiding in a basement. But the atmosphere is there, filling every space on Earth.

An atmosphere is the layer of gases that surrounds a planet or **star**. All planets and stars have an atmosphere. The Sun's atmosphere is hydrogen. Mars has a thin atmosphere of **carbon dioxide (CO_2)** with a

Although air itself is invisible, we can see, hear, and feel its effects as wind. For example, when clouds drift across the sky, leaves rustle, or your face is stung by a cold, windy blast.

bit of **nitrogen** and a trace of **water vapor**. Mercury has almost no atmosphere at all. Venus, almost the same size as Earth, has an atmosphere of carbon dioxide and nitrogen. It would be toxic to a visiting human. Each planet is surrounded by its own mixture of gases.

Earth's atmosphere is composed of a mixture of gases we call air. Air is mostly nitrogen (78 percent) and **oxygen** (21 percent), with some argon, carbon dioxide, **ozone**, water vapor, and other gases.

Take Note

Sketch a pie chart in your notebook to represent the percentage of gases in the atmosphere.

Permanent Gases

Nitrogen is the most abundant gas in our atmosphere. It is a stable gas, which means it doesn't react easily with other substances. When we breathe air, the nitrogen goes into our lungs and then back out unchanged. We do not need nitrogen gas to survive, but it does not harm us either.

Oxygen is the second most-abundant gas. It makes up about 21 percent of the air's volume. Because the oxygen atom is larger than the nitrogen atom, it accounts for 23 percent of air's **mass**. Oxygen is colorless, odorless, and tasteless. It is the most plentiful element in the rocks of Earth's crust. Oxygen combines with hydrogen to form water.

Oxygen and nitrogen are called **permanent gases**. The amount of oxygen and nitrogen in the atmosphere stays constant. Most of the other gases in the atmosphere table are also permanent gases, but are found in much smaller quantities.

Gases of the Atmosphere	
Gas	Percentage by volume
Nitrogen	78.08
Oxygen	20.95
Argon	0.93
Water vapor*	0.25
Carbon dioxide*	0.039
Ozone*	0.01
Neon	0.002
Helium	0.0005
Krypton	0.0001
Hydrogen	0.00005
Xenon	0.000009

*Variable gas

Many power plants burn coal to produce electricity. This can pollute the air with contaminants and adds carbon dioxide, a greenhouse gas, to the atmosphere.

Variable Gases

Air also contains **variable gases**. The amount of a variable gas changes in response to activities in the environment. Each of the variable gases listed in the table is also a **greenhouse gas**, which means it traps **thermal energy**, heating the atmosphere. Greenhouse gases are important for life on Earth. If we didn't have greenhouse gases, the Earth would be too hot during the day and too cold during the night to support life. If the amount of greenhouse gases increases, the atmosphere heats more.

Water vapor is the most-abundant variable gas. It makes up about 0.25 percent of the atmosphere's volume. The amount of water vapor in the atmosphere changes constantly. Water cycles between Earth's surface and the atmosphere through evaporation, condensation, and precipitation. You can get a feeling for the changes in atmospheric water vapor by observing clouds and noting the stickiness you feel on humid days.

Carbon dioxide is another important variable gas. It makes up only about 0.04 percent of the atmosphere. You can't see or feel changes in the amount of carbon dioxide in the atmosphere.

Investigation 1: What Is Weather?

Life Interacts with Air

Carbon dioxide plays an important role in the lives of plants and algae. These organisms remove carbon dioxide from the air during **photosynthesis**. Plants and algae convert light energy into chemical energy by making sugar (food) out of carbon dioxide and water. This process releases oxygen into the atmosphere. When living organisms use the energy in food to stay alive, they remove oxygen from the air and release carbon dioxide into the air.

There are other gases that you may have heard about. Ozone is a variable gas. It is a form of oxygen that accumulates in the **stratosphere**. Ozone is absolutely essential to life on Earth because it **absorbs** deadly ultraviolet radiation from the Sun. But ozone in high concentration can cause lung damage. In the lower atmosphere, ozone is an air pollutant.

Methane is a variable gas that is increasing in concentration in the atmosphere. Scientists are trying to figure out why this is happening. They suspect several things. Cattle produce methane in their digestive processes. Methane also escapes from coal mines, oil wells, and gas pipelines. It is a by-product of rice cultivation. Methane, like other greenhouse gases, absorbs energy coming up from Earth's surface.

These gases and a few other trace gases are all mixed together. Any sample of air is a mixture of all of them. If you rise higher in the atmosphere, there are fewer particles, but the ratio of the gases stays the same. The mixing is caused by constant movement of the air in the atmosphere near Earth's surface. Above about 90 kilometers (km), there is much less mixing. Very light gases (hydrogen and helium, in particular) are more abundant above that level.

Think Questions

1. What is the difference between permanent gases and variable gases in the atmosphere?
2. During daylight hours, plants and algae take in carbon dioxide and release oxygen. If humans remove forests, what might happen to the balance between these gases?

What's the connection between cows and climate change? As humans consume more beef, they raise more cattle, and cows produce methane. After carbon dioxide, methane is the second most abundant greenhouse gas in the atmosphere.

Investigation 1: What Is Weather?

A Thin Blue Veil

Not too hot, not too cold . . . just the right temperature. Our atmosphere is essential for life on Earth to exist.

It is cold in deep space. The temperature is in the neighborhood of –270 degrees Celsius (°C). That's nearly 200°C colder than it has ever been on Earth. Near stars, like the Sun, it is terrifically hot, reaching thousands of degrees. But there are places here and there in the universe where the temperature is between the extremes. These areas where life can exist are sometimes called the "Goldilocks Zone." Earth is one of those places. The average temperature on Earth is not too hot and not too cold, just right for supporting life.

Earth's deserts may be the hottest places on our planet, but plant and animal life still thrives in these harsh environments.

On any day, the temperature range on Earth is about 100°C. It ranges from about 45°C in the hottest place to –55°C at one of the poles. The extremes are 57°C in Furnace Creek Ranch, Death Valley National Park, California, recorded on October 7, 1919, and –89°C in Vostok, Antarctica, on July 21, 1983. So Earth's range of temperature is 146°C.

The frigid polar regions—the Arctic and the Antarctic—are the coldest places on Earth.

The Atmosphere

It is not only Earth's distance from the Sun that affects its temperature. Earth's atmosphere keeps the temperature within a narrow range that is suitable for life.

From space, Earth's atmosphere looks like a thin blue veil. Some people call it an ocean of air. The depth of this "ocean" is about 600 kilometers (km). The atmosphere is most dense right at the bottom, where it rests on Earth's surface. The air gets thinner and thinner (less dense) as you move away from Earth's surface, until it disappears.

Imagine a column of air that starts on Earth's surface and extends up 600 km to the top of the atmosphere. Scientists have discovered several layers in this column of air. Each layer has a different temperature.

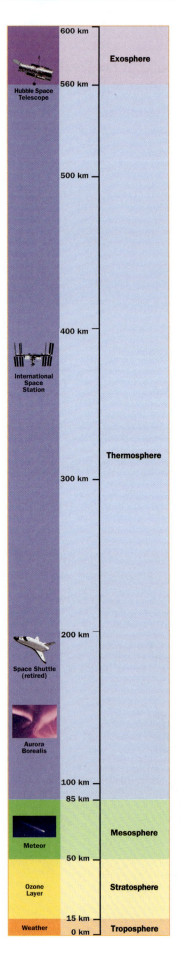

Earth's atmosphere

From space, Earth's atmosphere looks like a thin blue line around the planet.

The Troposphere

The layer of the atmosphere we live in is the **troposphere**. It starts at Earth's surface and extends upward for 10–24 km. Its thickness depends on the **season** and where you are on Earth. Over the warm equator, the troposphere is a little thicker than it is over the polar regions, where the air is colder. It also thickens during the summer and thins during the winter. A good average thickness for the troposphere is 15 km.

This bottom layer contains most of the organisms, dust, water vapor, and clouds found in the entire atmosphere. It also contains most of the air. And most important, weather happens in the troposphere.

The troposphere is where the action is. It is where differences in air temperature, humidity (moisture), **air pressure**, and wind occur. These properties are called **weather factors**. Meteorologists launch weather balloons twice a day to monitor worldwide weather factors. The balloons float up through the atmosphere to about 23 km. Weather factors will be investigated in detail as we continue to study weather.

Weather balloons carry instruments that monitor temperature, humidity, air pressure, and wind.

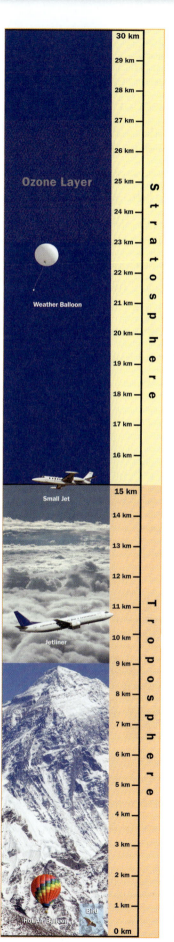

The first 30 km of the atmosphere

The Thinnest Layer

The troposphere is the thinnest layer, only about 2 percent of the depth of the atmosphere. It is also the most dense layer. It contains 80 percent of the total mass of the atmosphere.

Earth's surface (land and water) absorbs energy from the Sun and warms the air above it. Because air in the troposphere is heated mostly by Earth's surface, the air is warmest close to the ground. The air temperature drops as you go higher. At its upper limit, the temperature of the troposphere is about −60°C. The average temperature of the troposphere is about 25°C.

The Stratosphere

The stratosphere is the layer above the troposphere. It is 15–50 km above Earth's surface and contains almost no moisture or dust. It contains a layer of ozone, a form of oxygen. Ozone absorbs high-energy ultraviolet radiation from the Sun. The temperature stays cold until you reach the upper reaches of the stratosphere. There, energy absorption by ozone warms the air to about 0°C.

The **jet stream** is a fast-flowing river of wind between the lower stratosphere and the upper troposphere. It flows generally west to east in the region. Many military and commercial aircraft take advantage of the jet stream when flying from west to east.

Above 8,000 m, climbers on Mount Everest enter what is known as the "death zone." No human can survive long there due to the lack of oxygen in the thin air. Climbers usually bring oxygen along to help them survive the summit.

Did You Know?

Mount Everest, located in Nepal and Tibet, is the highest landform on Earth, rising 8.8 km into the troposphere. The air temperature at the top of the mountain is well below freezing most of the time.

Charged particles in the upper atmosphere can produce eerie light shows above Earth's magnetic poles. In the Northern Hemisphere, these shimmering curtains of color are called the aurora borealis, or the northern lights.

The Mesosphere

The **mesosphere** is above the stratosphere, 50–85 km above Earth's surface. The atmosphere reaches its coldest temperature, around –90°C, in the upper mesosphere. The mesosphere is where meteors burn up while entering Earth's atmosphere, producing shooting stars.

The Thermosphere

Beyond the mesosphere, 85–560 km above Earth, is the **thermosphere**. The thermosphere is the least-understood layer of the atmosphere and the most difficult to measure. The air is extremely thin. The thermosphere is the region of the atmosphere that is first heated by the Sun.

A small amount of absorbed **solar energy** can produce a large temperature change. When the Sun is extra active with sunspots or solar flares, the temperature of the thermosphere can rise to 1,500°C or higher!

The thermosphere has regions that contain a large number of electrically charged ions. Ions form when intense radiation from the Sun hits particles in the atmosphere. These ions produce the colorful aurora borealis and the aurora australis, the northern and southern lights.

Investigation 1: *What Is Weather?*

The International Space Station and some other satellites, like the Hubble Space Telescope, orbit in the thermosphere.

Atmospheric Temperatures

These four layers (troposphere, stratosphere, mesosphere, and thermosphere) are separated by temperature. They have no sharp boundaries or abrupt changes in gas composition. As average temperatures change with the seasons, the boundaries between layers move up or down a little.

Beyond the thermosphere, Earth's atmosphere gradually becomes space. This area is the **exosphere**, where particles from the atmosphere escape into space. It extends from the top of the thermosphere up to 10,000 km. In this region, the temperature plunges to the −270°C of outer space. The concentration of atmospheric gases fades to nothing.

Pressure

The 600 km column of air that makes up the atmosphere pushes down on the surface of Earth with a lot of force. We call this force air pressure, or atmospheric pressure. We are not aware of all that pressure because we are adapted to live under it. A mass of about 1 kilogram (kg) pushes down on every square centimeter of surface on Earth.

Here is another way to look at it. If all the mass of the air were replaced with solid gold, the entire planet would be covered by a layer of gold a little more than half a meter deep. That's a lot of gold, but the atmosphere is much more valuable.

Think Questions

1. How is Earth's atmosphere like the ocean? How is it unlike the ocean?
2. Why do you think airplanes do not fly high in the stratosphere?

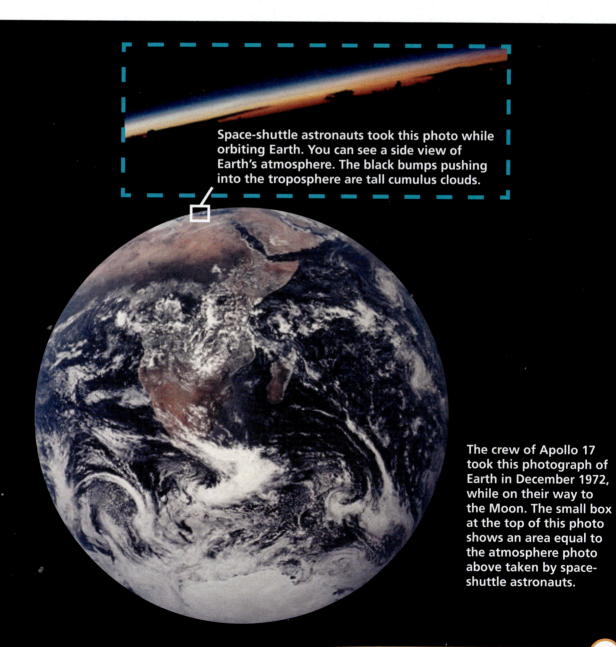

Space-shuttle astronauts took this photo while orbiting Earth. You can see a side view of Earth's atmosphere. The black bumps pushing into the troposphere are tall cumulus clouds.

The crew of Apollo 17 took this photograph of Earth in December 1972, while on their way to the Moon. The small box at the top of this photo shows an area equal to the atmosphere photo above taken by space-shuttle astronauts.

Investigation 1: What Is Weather?

What Is Air Pressure?

We are all under pressure from the weight of the atmosphere. Why don't we feel it all the time? Why do your ears pop driving over a mountain pass? The answers have to do with air pressure.

The atmosphere is composed of air. Air has mass. In fact, a column of air 1 square centimeter (cm^2) in area reaching to the top of the atmosphere has a mass of 1.2 kilograms (kg). Say the top of your head has a surface area of 150 cm^2. That means every time you stand under the open sky, you have 180 kg of air pushing down on your head! That is like wearing a hat with a refrigerator on it. Is it safe to go outside?

Yes, it is. The force applied by the air above you is called atmospheric pressure. Life on Earth has evolved in this high-pressure environment, so we can handle the pressure just fine. In fact, most of the time we are totally unaware of the pressure.

Air pressure decreases as elevation increases. As this climber hikes higher, the air around him is less dense.

We do feel the air pressure when it changes quickly. Have you ever traveled to a mountain and noticed a popping in your ears as you go up or down the mountain? Sometimes this happens in an airplane when it changes altitude rapidly, or in a car going up and down a mountain road. Keep that experience in mind as you think more about atmospheric pressure.

Sometimes you can see evidence of change in pressure even if you cannot feel it. In an airplane, you might notice that tightly sealed packets of peanuts or chips are puffed up like balloons.

This bottle of air looked normal when it was closed tightly at high altitude. Then it came down to sea level. Its crushed condition is evidence of changed air pressure.

The plastic bottle crumpled when the outside pressure became greater than the inside pressure.

Investigation 2: Air Pressure and Wind

What Causes Air Pressure?

Air particles have mass, so they are pulled to Earth by gravity. The air surrounding Earth has weight. Atmospheric pressure is the weight of the air pushing on Earth's surface and on everything near Earth's surface.

Remember, air particles are zipping around individually. So what prevents gravity from attracting them all to the ground? Why aren't we walking around knee-deep in a thick soup of oxygen and nitrogen particles?

The answer is **kinetic energy**. Kinetic energy is energy of motion. The gas particles in air have so much energy of motion that they are bouncing around each other in all directions. Air resists being compressed because of this constant banging of air particles into one another at an astounding rate of several billions of times per second.

Here is an important point: The atmospheric pressure is not the only thing pushing down on Earth. The particles are also banging around and colliding with the air particles above and at their sides. As a result, atmospheric pressure pushes with equal force in every direction.

As these balloons and their passengers float higher, they find themselves above much of the air. So the atmospheric pressure on them is lower.

Variation in Pressure

Atmospheric pressure is not the same everywhere. Air pressure is caused by the mass of air being pulled to Earth. So what happens if you go up into the atmosphere, high above Earth's surface, where there is more air below you and less air above you?

If you went for a ride in a hot-air balloon, you might find yourself 2 kilometers (km) above the land. Up there, 2 km of air is below you, so that 2 km of air is not applying pressure where you are. Therefore, the atmospheric pressure is less at that altitude.

Because we live at the bottom of a sea of compressible air, the atmosphere is most dense at Earth's surface. At the surface, with the greatest amount of air overhead, we have the greatest pressure. The air particles are pushed close together because gases can be compressed. And when more particles are present in a given volume, the gas is more dense.

Did You Know?

The earliest known hot air balloons did not hold passengers, but were used for signaling. These balloons may have been used as early as third century BCE.

As you go up in the atmosphere, pressure decreases. The air expands because less force is pushing the particles together. The air gets less dense. Mount Everest is over 8 km high. At the peak, the atmospheric pressure is only one-third the pressure and **density** that it is at sea level. Have you ever seen pictures of climbers laboring up the highest reaches of the mountain? Most of them are using oxygen supplies. Because there is only one-third as much air at that elevation, a breath supplies one-third as much oxygen to the body. It is rare for a climber to reach the summit without extra oxygen.

Let's return to those ears popping when you travel over the mountain. What does it have to do with variation in atmospheric pressure?

Measuring Air Pressure

The air pressure that meteorologists talk about on the news is caused by the mass of air pushing on a certain point on Earth. Elevation is one factor that causes pressure to vary. A number of other factors also affect air pressure.

Air Pressure and Elevation

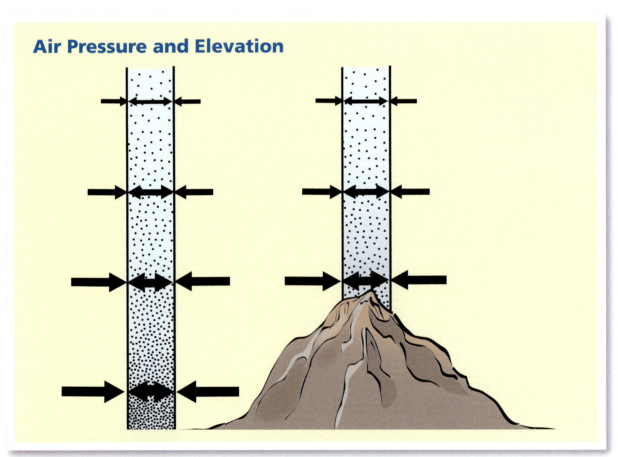

Air pressure on top of a mountain is less than it is at sea level.

Torricelli proposed that the air in our atmosphere has weight and exerts a force by pushing down. The device he invented to measure this force is a barometer, from the Greek word *baros*, meaning "weight."

Meteorologists use a **barometer** to measure air pressure. An Italian naturalist named Evangelista Torricelli (1608–1647) invented the first barometer in 1643. He filled a long glass tube with mercury and turned it upside down in a dish also filled with mercury. A small amount of the mercury in the tube ran out of the tube and into the dish, leaving an empty space above the mercury in the tube. This space was a **vacuum**. A vacuum is a space containing almost no **matter**, not even air.

What held the heavy column of mercury up in the tube? Air pressure pushed down on the mercury in the dish. The pressure was distributed throughout the mercury, including the mercury in the tube.

As Torricelli closely observed his column of mercury, he noted that the level of mercury moved up and down a little from day to day. He reasoned that the level of mercury changed because of changes in atmospheric pressure. As atmospheric pressure increased, there was more pressure on the mercury. Therefore, it rose farther up the tube. Torricelli had invented the first barometer, an instrument for observing and measuring changes in atmospheric pressure.

Did You Know?

A column of atmosphere with an area of 1 cm^2 has a mass of 1.2 kg. If a column of mercury has a cross section of 1 cm^2, it has a mass of 1.2 kg. So the force from air pressure is exactly balanced by the weight of the mercury in the tube.

Take Note

Review your notes about the closed-system pressure jar. How is Torricelli's open-system barometer different from the setup you used in class? How is it similar?

This aneroid barometer dial shows air pressure in both inches and millibars. Compare the previous reading (brass needle) to the current reading (black needle). Is air pressure rising or falling?

Modern Barometers

Today, meteorologists often use aneroid barometers. These barometers are much smaller and more portable than mercury barometers.

At the heart of an aneroid barometer is a sealed chamber with a spring inside. The chamber acts like a bellows. All the air is removed from inside the bellows. As atmospheric pressure increases, air particles bump into the bellows more frequently. This compresses the bellows, and the sealed chamber becomes shorter. When pressure is lower, air particles hit the bellows less frequently. The bellows expands. A pointer attached to the bellows moves along a scale to show the change in pressure.

Modern barometers are digital. They use electronic pressure sensors in circuits that produce digital displays of atmospheric pressure.

Air Pressure at High Altitude

Weather balloons have an electronic sensor and a transmitter attached to a barometer. Information about air pressure is sent from a weather balloon to a receiving unit on Earth. That is how we collect information about air pressure at high altitudes.

Imagine you are going on vacation. When you packed your suitcase at home and included a half-filled bottle of lotion, the air pressure inside the bottle was equal to the air pressure on the ground. As it rides in the airplane up to the top of the troposphere, 80 percent of the atmosphere is now below the plane. Most passenger airplanes have internal air pressure similar to 2,000 m above ground. So, the pressure inside the bottle is greater than outside. When you open the bottle, look out! Air will rush out and is likely to push lotion with it. A sealed bottle with air in it is acting like a simple barometer.

Inside a Barometer

Changing air pressure causes a bellows to compress and expand inside an aneroid barometer.

Take Note

Review your work with the syringe. What might happen to air in a syringe if you took a clamped-shut syringe of air on an airplane?

Twice every day, meteorologists launch weather balloons at about 800 locations around the world. During their 2-hour flights, they gather valuable atmospheric data for weather forecasts and research.

Investigation 2: *Air Pressure and Wind*

Pressure Units

Scientists and meteorologists use several units to measure and report pressure. For historical reasons, inches or centimeters of mercury are used in science experiments. Meteorologists, on the other hand, look at the average atmospheric pressure at sea level and call that 1 bar. If you are relaxing at the beach, the pressure around you will be 1 bar, or close to it.

The bar has been subdivided into 1,000 equal millibars (mb). You used these units to record the local atmospheric pressure on the class weather chart. In practice, standard pressure is actually 1,013 mb. Any pressure below 1,013 mb is lower than normal pressure, and over 1,013 mb is higher than normal pressure.

Think Questions

1. When you drive down a mountain, what makes your ears experience those interesting and sometimes uncomfortable sensations?
2. Why doesn't air pressure crush an empty soft-drink can as you drive down a mountain?
3. Why doesn't air pressure crush an empty soft-drink can that is sitting on a table in your house?
4. A meteorologist says that the air pressure is getting lower. What would you expect to see happen to Torricelli's mercury barometer?
5. If Torricelli had drilled a little hole at the top of the glass tube holding his mercury column, what would have happened to his barometer?

Good beach weather typically means atmospheric pressure around 1 bar, or 1,000 millibars. If pressure begins to fall rapidly, you may wish to head home—a storm could be coming.

Unlike a helium balloon, which is filled with a lighter-than-air gas, a hot air balloon is filled with heated air. Heating a gas causes its particles to move faster and push farther apart, becoming less dense.

Density

A hot air balloon is a beautiful sight to see floating across the morning sky. The balloon drifts away gracefully. It appears so light, yet it is carrying six people in its basket. How does the balloon pilot lift the system into the sky and safely lower it back to the ground? The answer has to do with density.

Imagine you have a package of regular rubber balloons. Fill one with water, tie it off, and give it to a friend. Fill a second, identical balloon with air until it is the same size as the water balloon. Tie it off and give it to a second friend to hold. Fill a third balloon with helium, same size as the other two, and tie it off.

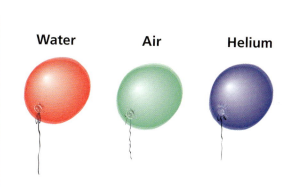

Review the three identical balloons, all filled to exactly the same volume, each tied off so nothing can get in or out. What's different? The kind of material in the three balloons. Ready to try a little experiment?

You and your friends hold the balloons at the same height above the floor. On the count of three, you will all release your balloons and observe what happens. Three … two … one … go!

The water balloon will plunge to the floor, the air balloon will drift slowly to the floor, and the helium balloon will float up to the ceiling. Why?

Matter and Mass

It comes down to how much stuff there is in each balloon, relative to the stuff outside. The scientific word for stuff is matter. The amount of matter in an object is its mass. Matter is made of particles called atoms. So the mass of an object depends on *how many* atoms there are in the object and the mass of the atoms.

The atoms in solids (glass, steel) and liquids (water, rubbing alcohol) are packed together as close as they can get. This means that a volume of water has lots of atoms. That makes water pretty heavy.

In gases, the atoms are not packed together as close as they can get. There is a lot of space between atoms. Remember how you could compress air in the syringe? Air and helium are gases, so they are pretty light.

The nitrogen and oxygen atoms in the air are fairly large, but helium atoms are small. Generally speaking, small atoms weigh less than large atoms. So a volume of helium is much lighter than an equal volume of air.

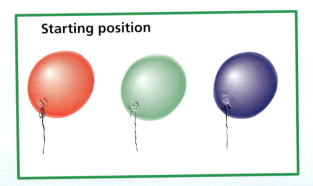

Starting position

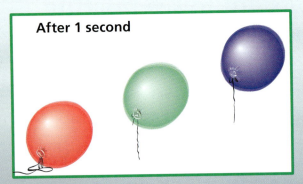

After 1 second

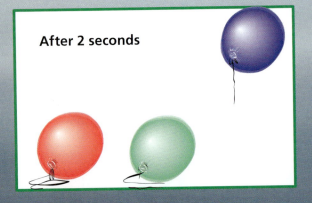

After 2 seconds

Calculating Density

The amount of matter in a volume of material determines the material's density. Density is defined as mass per volume of an object. When you have equal volumes of a number of different materials, you can find out which one is most dense and which one is least dense by weighing them. The heaviest one is the most dense. The lightest one is the least dense.

The important idea in this discussion is that you need to compare the mass of *equal volumes* of different materials to determine which one is most dense.

The usual way of stating density is mass per volume. The word per means "divided by." Density can be written as an equation.

$$\text{Density} = \frac{\text{Mass}}{\text{Volume}}$$

Density equals mass divided by volume. If you remember "mass per volume," you will always know how to set up your equation when it comes time to calculate density.

This sealed bottle with a note (and air) inside bobs on the water's surface. Under what conditions might a glass bottle sink instead of float?

Objects sink or float in water depending on their density. Density is how much mass a certain volume of a substance has. A marshmallow floats while a marshmallow-sized rock sinks.

Density Interactions

One way to determine the density of a material is to determine the mass of a 1 cubic centimeter (cm^3) sample of that material. If there is a lot of matter in a cubic centimeter sample of a material, it is dense. If there is very little matter in a cubic centimeter sample, it is not very dense.

One milliliter (mL) is exactly the same volume as 1 cm^3. Liquids are generally measured in milliliters, and solids and gases are measured in cubic centimeters.

Mass is measured in grams (g). Pure water has a density of 1 g/mL. Materials with densities greater than 1 g/mL (or 1 g/cm^3) are more dense than water. Materials with densities less than 1 g/mL are less dense than water.

What will happen if you put a rock with a density of 3 g/cm^3 in water? It will sink like, well, a rock. What about if you put a cork with a density of 0.45 g/cm^3 in the water? Yes, it will float because it is less dense than water.

Materials that are more dense than water sink in water. Materials that are less dense than water float in water. That is the way it always works.

Density of Gases

Remember the balloons? We know the density of water, but what about the air and the helium?

Material	Density
Water	1.0 g/mL
Air	0.0013 g/mL
Helium	0.0002 g/mL

The table shows that air and helium are not very dense. There is very little mass in a cubic centimeter of gas. When the three balloons were released, the dense water-filled balloon pushed through the less-dense air surrounding it. It sank straight to the floor.

The air-filled balloon was almost the same density as the surrounding air. Rubber is more dense than air, and the air inside was compressed a little by the stretchy balloon, making the air-filled balloon a little more dense than the surrounding air. It sank slowly to the floor.

The helium-filled balloon was quite a bit less dense than the surrounding air. Just like a cork in water, the less-dense helium balloon floated as high as it could.

If you let go of the ribbons holding these helium-filled balloons, you'll put a few more stars in the sky. The helium gas inside the balloons is less dense than air, so the balloons float.

Investigation 3: Convection

The sky may look calm and still, but the gas particles that make up air are in constant motion. How fast? That depends on the air temperature.

Density of Air

Air is gas. The particles in gases are not in contact with one another. Gas particles move around freely in space.

When energy transfers to matter, the kinetic energy (movement) of the particles increases. The increased motion causes matter to expand. When matter expands, the particles do not get bigger. They get farther apart. This is a very important point. It is the distance between particles that increases, not the size of the particles.

When matter expands, the particles get farther apart. What do you think that does to the density of the material? The density gets lower. When the particles get farther apart, each cubic centimeter has fewer particles. Fewer particles in a volume means lower density.

This is a general rule of matter. When matter gets hot, it expands, and the density goes down. The idea of density will be an important concept in our investigation of weather and the processes that cause it.

Think Questions

1. What do you think the density of a person might be? Explain.
2. Why do you think hot-air balloons can rise into the air? How do the pilots get their hot-air balloons back to Earth?
3. Weather happens in the atmosphere. What do you think happens when some air heats up and other air is cool? What weather might be the result?

Density with Dey

Mr. Dey's students had been working with salt solutions. When salt dissolves in water, it forms a solution.

The students found out that the more salt they dissolved in a volume of water, the more dense the solution became. Mr. Dey's students received three salt solutions from another class and were challenged to put them in order by density.

Group 1 put 25 milliliters (mL) of the blue solution on a scale and found that it had a mass of 25 grams (g). They measured 25 mL of green solution into another cup. Its mass was 30 g.

The students announced, "We weighed equal volumes of two solutions. The green solution is heavier, so it is more dense. It has more mass per volume than the blue solution."

Group 2 put 25 mL of blue solution on a scale and found that it had a mass of 25 g. But then they made a little mistake. They put 50 mL of yellow solution in a cup and found that it had a mass of 55 g.

These students measure and record observations carefully so their experiments can be repeated by others, with similar results.

Investigation 3: Convection

When they realized what they had done, Reggie said, "Uh-oh, we didn't measure equal volumes. We have to start over."

"Maybe not," said Yolanda. "We weighed twice as much yellow solution as we should have. If we had used half as much, it would have weighed half as much. All we have to do is divide the mass by two to find out the mass of 25 mL of yellow solution."

They did the math and found that 25 mL of yellow solution had a mass of 27.5 g.

The two groups put their data into a table.

Solution	Volume	Mass
Blue	25 mL	25 g
Green	25 mL	30 g
Yellow	25 mL	27.5 g

The students could now easily compare equal volumes of the three solutions to see which one was heaviest and, therefore, the most dense. They determined that green was most dense, blue least dense, and yellow in the middle.

Mr. Dey had a question. "What is the mass of 1 mL of each of the solutions?"

Reggie offered, "25 mL of blue solution has a mass of 25 g, so 1 mL of blue has a mass of 1 g."

"And the green solution?" asked Mr. Dey.

Reggie's group thought about it this way.
- Twenty-five milliliters of green has a mass of 30 g. That is more than 1 g for each milliliter.
- They drew a pie chart to help them think about the problem. Each slice of the pie represented 1 mL, or 1/25 of the total volume.
- One milliliter is 1/25 of the total volume. Each milliliter must have 1/25 of the total mass.
- They divided 30 g by 25 mL to find the mass of 1 mL of green solution.

The students discovered the definition of density.

Density can be written as an equation.

$$\text{Density} = \frac{\text{Mass}}{\text{Volume}} = \frac{g}{mL}$$

The equation can be used to calculate the density of the green solution.

$$\text{Density} = \frac{\text{Mass}}{\text{Volume}} = \frac{30\ g}{25\ mL} = 1.2\ g/mL$$

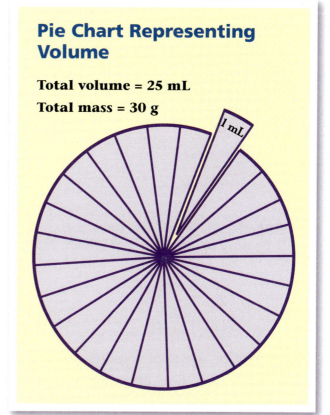

Pie Chart Representing Volume

Total volume = 25 mL
Total mass = 30 g

Each slice of the pie represents 1 mL, or 1/25 of the total volume. Each milliliter must have 1/25 of the total mass.

The different colors help us tell the solutions apart. But the different-shaped containers make it difficult to use volumes. Why is it important to compare the same volume each time?

The liquids in this test tube have different densities. The less-dense yellow liquid is floating on the more-dense blue liquid.

Density equals mass divided by volume. If you remember "mass per volume," you will always know how to set up your equation when it comes time to calculate a density.

What is the density of the yellow solution? The original mass of the yellow solution was 55 g, and the volume was 50 mL.

$$\text{Density} = \frac{\text{Mass}}{\text{Volume}} = \frac{55 \text{ g}}{50 \text{ mL}} = 1.1 \text{ g/mL}$$

Take Note

If you carefully pour a little bit of each of the colored solutions from Mr. Dey's class into a vial, what do you think would happen?

Solution	Density
Blue	1.0 g/mL
Yellow	1.1 g/mL
Green	1.2 g/mL

Can you describe the result? Record in your notebook your ideas about what would happen. Include a diagram to help you explain.

Convection

It's time to cook dinner. You fill a pan with cool water, place it on the stove, and turn on the burner. After about 10 minutes, it's time to cook. The burner only touched the bottom of the pan, so how did the water heat up?

Considering a different heat source may help answer the question. If you put a couple of centimeters of water in a metal pie pan and support it over a candle, you can slowly heat the water. The energy from the flame will heat a small area of the pan and the water directly over the flame will warm up.

At first, only the center of the pan and water directly above the flame get hot.

Stovetop cooking in liquid works because of convection, the transfer of energy by the movement of a fluid, like water for boiling a pot of pasta.

Investigation 3: *Convection*

A small mass of water will heat up. The question is, what happens to that hot water?

The water expands as the water particles gain kinetic energy and push farther apart. The expanding water is less dense than the surrounding water. The warm water floats up.

Next, the small amount of heated water moves upward.

When the warm water reaches the surface, it flows outward. Water at the surface cools because it is no longer near the heat source and energy can transfer away from the water to the environment.

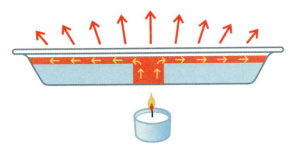

The heated water spreads out over the surface and transfers energy to the air above the pan.

When the water reaches the edge of the pan, the water has cooled down and is more dense. It flows down the sides of the pan, across the bottom, and back toward the center of the pan. As the water nears the hot metal, it begins to warm again. The hot water floats up to repeat the cycle.

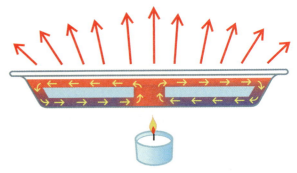

Once cooled, the cooler water sinks down the sides and along the bottom toward the heat source, where the water is reheated and the convection cycle continues.

The flow of water in the pan, driven by a localized heat source, is **convection**.

Convection happens only in **fluids**. Gases and liquids are both fluids because they flow. Fluid near an energy source heats up and expands. The hot fluid becomes less dense and floats upwards. The energy in the particles of hot fluid is carried to a new location. As the fluid cools down (energy transfers away from the fluid), it cools and contracts, making it more dense. The cool fluid flows downward again. The mass of fluid flowing in a cycle is called a **convection cell**.

Air is a fluid. Energy can transfer to air, causing it to expand. When air expands, it becomes less dense. Less-dense air will float up in the atmosphere. In the next few investigations, we will see how the process of convection moves air around the planet and affects all types of wind, from gentle breezes to powerful, dangerous storms.

What makes winter wintry? In winter, we have fewer hours of sunlight and less direct sunlight. Less solar energy results in lower temperatures.

Seasons

What do you imagine when you read these words: summer, fall, winter, spring?

Most of us come up with a mental picture or two. Summer means shorts, sandals, and T-shirts, swimming, and fresh fruits and vegetables. Winter means heavy coats and short days, perhaps with a blanket of snow on everything.

Seasons are pretty easy to tell apart in most parts of the United States. The amount of daylight, precipitation, the average temperature, and the behaviors of plants and animals are a few familiar indicators of the season. But what causes the predictable change of season? What have you learned in class that helps explain the reasons for the seasons?

Investigation 4: *Radiation* 53

 North Star

As Earth Tilts

Let's start with a quick review of some basic information about Earth.

Earth spins on an imaginary axle called an **axis**. The axis passes through the North and South Poles. This spinning on an axis is called **rotation**. It takes 24 hours for Earth to make one complete rotation.

Earth travels around (**orbits**) the Sun. Traveling around something is called **revolution**. One revolution of Earth around the Sun takes 365 and 1/4 days, which we call 1 year.

> **Take Note**
>
> Record the reasons for seasons on Earth. You can add more after reading this article, but record your first ideas now.

Earth isn't oriented straight up and down on its axis as it revolves around the Sun. It is tilted at a 23.5° angle.

The average distance between the Sun and Earth is about 150 million kilometers (km). Earth's orbit is slightly oval, so during its revolution around the Sun it is sometimes a bit farther away from and sometimes a bit closer to the Sun. This distance is so insignificant that it is not a contributing factor to the cause of the seasons.

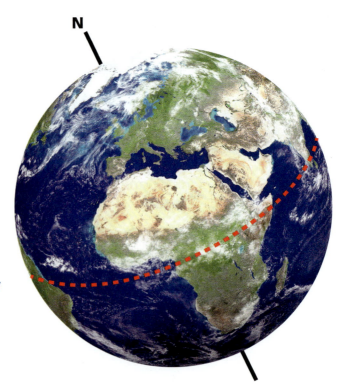

As Earth completes each 24-hour rotation and each 365 and 1/4-day revolution, its axis always points in the same direction—tilted at a 23.5° angle.

54

Autumn leaves turning red and orange are how some plants respond to the decreasing daylight as summer ends and winter approaches.

It would seem logical that summer would be when Earth is closest to the Sun. That idea is wrong. Each year when Earth is closest to the Sun, the Northern Hemisphere experiences winter. The reasons for the seasons are linked to Earth's tilt, not to the distance from the Sun.

Think about Earth orbiting the Sun. As Earth orbits, it also rotates on its axis, one rotation every 24 hours. Here is something important: Earth's North Pole points toward a reference star called the **North Star**. No matter where Earth is in its orbit around the Sun, the North Pole points toward the North Star, day and night, every day all year.

Take Note

Is Earth closer to the Sun when it is winter or summer in the United States? Is distance from the Sun a reason for seasons on Earth?

Investigation 4: *Radiation*

North Star

In our Northern Hemisphere winter, when the North Pole is tilted away from the Sun, even at noon the Sun is low in the sky and sunlight strikes Earth at an indirect angle. In this region, solar energy—heat and light—is spread out across a larger area of Earth's surface.

Tilt Equals Season

Look at the illustration that shows where Earth is in its orbit around the Sun at each season. Notice that the North Pole points toward the North Star in all four seasons.

Study the Earth diagram in the summer **solstice** position. Because of the tilt, the North Pole is leaning toward the Sun. When the North Pole is leaning toward the Sun, daylight is longer, and the angle at which light hits the Northern Hemisphere is more direct. Both of these factors result in more solar energy falling on the Northern Hemisphere. It is summer even though Earth is actually farther from the Sun. (And when it is summer in the Northern Hemisphere, it is winter in the Southern Hemisphere.)

Look at the position of Earth 6 months later (at winter solstice). Now the opposite is true. Even though Earth is closer to the Sun at this time, the Northern Hemisphere is tilted away from the Sun. Daylight hours

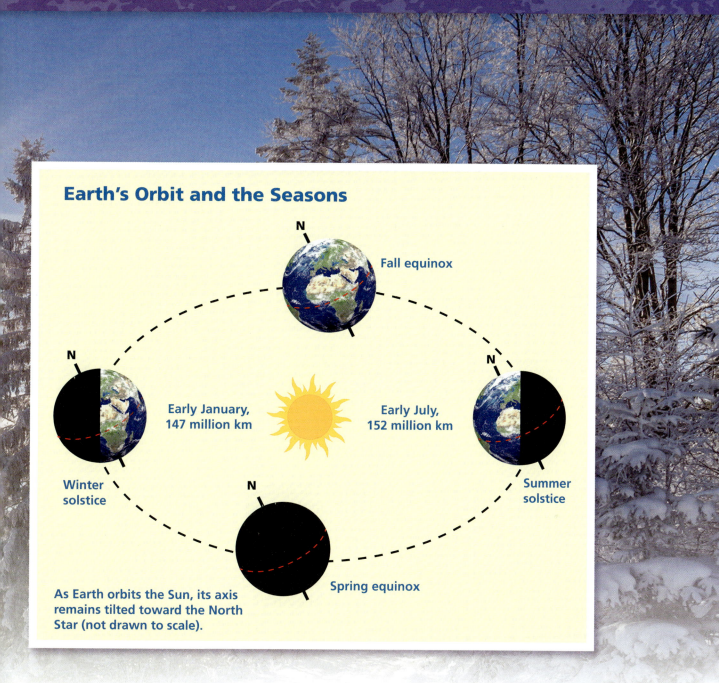

Earth's Orbit and the Seasons

Fall equinox

Early January, 147 million km

Early July, 152 million km

Winter solstice

Summer solstice

Spring equinox

As Earth orbits the Sun, its axis remains tilted toward the North Star (not drawn to scale).

are shorter, and sunlight does not come as directly to the Northern Hemisphere, so it gets less solar energy. It is winter in the Northern Hemisphere.

Four days in the year have names based on Earth's location around the Sun. In the Northern Hemisphere, summer solstice is June 21 or 22, when the North Pole tilts toward the Sun. Winter solstice is December 21 or 22, when the North Pole tilts away from the Sun.

The 2 days when the Sun's rays shine straight down on the equator are the **equinoxes**. On these 2 days, Earth's axis is tilted neither away from nor toward the Sun. *Equinox* means "equal night." Daylight and darkness are equal (or nearly equal) all over Earth. There are two equinoxes each year, the spring (also called vernal) equinox in March and the fall (also called autumnal) equinox in September.

Investigation 4: *Radiation*

Length of Daylight in the Northern Hemisphere

Latitude	Summer solstice	Winter solstice	Equinoxes
0° N	12 hr.	12 hr.	12 hr.
10° N	12 hr. 35 min.	11 hr. 25 min.	12 hr.
20° N	13 hr. 12 min.	10 hr. 48 min.	12 hr.
30° N	13 hr. 56 min.	10 hr. 4 min.	12 hr.
40° N	14 hr. 52 min.	9 hr. 8 min.	12 hr.
50° N	16 hr. 18 min.	7 hr. 42 min.	12 hr.
60° N	18 hr. 27 min.	5 hr. 33 min.	12 hr.
70° N	24 hr.	0 hr.	12 hr.
80° N	24 hr.	0 hr.	12 hr.
90° N	24 hr.	0 hr.	12 hr.

Alaska, our northernmost state, receives more sunlight in spring and summer than any other state. One-third of Alaska lies above the Arctic Circle, in the "land of the midnight sun."

Day and Night

We take day and night for granted. They always happen. Earth rotates on its axis, and the Sun appears to rise, then the Sun appears to set. This cycle has happened for much longer than humans have been on Earth. It will most likely continue for millions more years.

Because Earth is tilted, the length of day and night for any one place on Earth changes as the year passes. This table shows how hours of daylight change at different **latitudes** during the year. When it is summer in the Northern Hemisphere, the North Pole tilts toward the Sun. During this time, the Sun never sets at the North Pole. Above the Arctic Circle (66.5° north), daylight can last up to 24 hours in the summer. Darkness can last up to 24 hours during the winter.

Think Question

Go back to the notebook entry about the reasons for the seasons that you made at the beginning of this article. What do you need to add? What do you need to change?

In this simple thermometer, the blue liquid in the tube rises (expands) and falls (contracts) as the temperature around the bulb changes.

Thermometer: A Device to Measure Temperature

Is the oven ready for this pie? You look flushed, do you have a fever? The fish are not eating. Is it warm enough in the aquarium? The ice cream is soft. Is that freezer working?

To answer these questions, we reach for a **thermometer**. And there are many kinds to reach for.

How Thermometers Work

While thermometers may look different, all thermometers work the same way on a basic level. Some property of a material changes as it gets hot. You may already be familiar with several kinds of thermometers. The old standby is the glass tube filled with alcohol or mercury. This is how it works.

A thin, heat-tolerant, glass tube with a bulb at one end is partly filled with alcohol or mercury. The liquid extends partway up the tube. The tube is sealed and attached to a backing that has a scale written on it.

When the bulb touches something hot, the liquid inside expands. The volume of liquid increases. The added volume of liquid can only go up into the tube. The distance the liquid pushes up the tube indicates the temperature of the material touching the thermometer bulb.

Investigation 4: Radiation

Early Thermometers

The first closed-tube thermometer was invented by Ferdinand II de' Medici (1610–1670) in 1641. He used alcohol in the tube. During the 1700s, improvements to closed-tube thermometers made it possible to conduct experiments involving accurate temperature measurements.

In 1714, German physicist Daniel Gabriel Fahrenheit (1686–1736) made a mercury thermometer. He also developed the Fahrenheit temperature scale. On the Fahrenheit scale, 32 degrees Fahrenheit (°F) is the freezing point of water. The boiling point of water is 212°F. In 1742, Swedish astronomer and physicist Anders Celsius (1701–1744) devised a new temperature scale. Zero degrees Celsius (°C) is the freezing point of water, and 100°C is the boiling point of water. This used to be called the centigrade scale, which means "hundred steps." In 1948, it was renamed the Celsius scale, in honor of Anders Celsius.

A friend of Galileo, Ferdinand II de' Medici invented an early thermometer and set up the first international organization for temperature observations.

The temperature scale developed by Anders Celsius is used in most of the world and by scientists everywhere. In this scale, 100 degrees separate the freezing point (0°C) and boiling point (100°C) of water.

Did you ever wonder how your oven thermometer knows how hot it is inside? A bimetallic strip changes shape as the oven heats up and moves the pointer on the thermometer dial.

Modern Thermometers

There are many types of modern thermometers. Oven thermometers and some wall thermometers look a little like pocket watches. Inside is a **bimetallic strip**. Bimetallic strips are made of two metals stuck together. The two metals expand at different rates when they get hot. When the temperature increases, the black-colored part of the strip expands (lengthens) more than the other. The strip bends. A pointer attached to the bending metal strip points to the temperature.

Tropical-fish owners keep thermometers right in the aquarium. One kind is a thin, flat strip of plastic, like a piece of black plastic ribbon, that has liquid crystals inside. Liquid crystals change color within a very narrow temperature range. A liquid-crystal thermometer has a series of little pockets in the strip. Each pocket holds a different mix of liquid crystals to indicate a single temperature. You look at the strip to see which number is glowing. That is the temperature.

Bimetallic strip thermometers are often used in ovens and refrigerators because they both measure *and* control temperature.

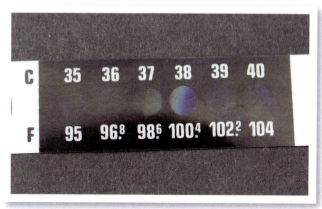

A liquid crystal thermometer uses heat-sensitive chemicals to indicate temperature.

Investigation 4: Radiation

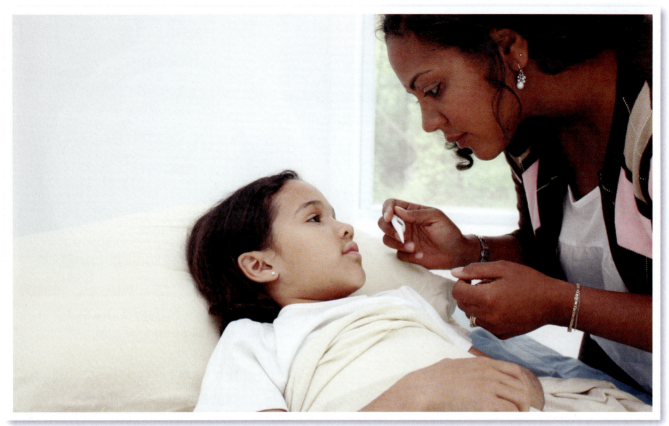

Why do doctors take your temperature? Body temperature can reveal a lot about your physical condition, including whether you have an infection. Normal body temperature is 37°C (98.6°F).

The last time you went to the doctor for a checkup, you may have had your temperature taken with a digital thermometer. These modern thermometers are very accurate and easy to use. You slip the probe end under your tongue for a few seconds. Inside the probe is a circuit with electricity flowing through it. Part of the circuit flows through a piece of wire that changes resistance as the temperature changes. When the electronic circuitry detects that the current flowing in the probe circuit has stabilized, that means the temperature is no longer changing. The electronic thermometer measures the amount of current flowing in the circuit and displays the temperature on a little digital screen.

That is just a small sampling of the many different thermometers found in common and specialized applications.

Think Questions

1. Where do you find thermometers in your daily life?
2. Why does the liquid rise and fall in the tube of a glass thermometer?

A digital thermometer uses electric circuitry to detect temperature and provides a numeric readout.

A Water-Density Thermometer

Galileo Galilei (1564–1642) invented one of the first thermometers in 1596. Galileo knew that water expands as it warms up. Warm, expanded water is less dense than cold water. He filled some small glass balls partway with colored water and sealed them shut. The balls of colored water all floated in water because of the light density of the glass ball systems. Galileo then attached a weight to each ball. The weights were adjusted to give each ball a slightly different mass and thus a different buoyancy.

When the water was cold (at its most dense), all the balls floated. As the water warmed up, becoming less dense, balls would sink.

By placing the balls in a column of water, with the lightest on the top, Galileo produced a thermometer. If all the balls were on the bottom of the cylinder, it was really hot.

Modern versions of the Galileo thermometer have temperatures printed on the weights. The number on the lowest floating ball shows the temperature of the system.

Each floating "bubble" carries a weight disk that gives that bubble a unique buoyancy. Therefore, each bubble indicates a different temperature.

The colorful water-density thermometer is used today mostly for decoration. Though not very precise, it is still fairly accurate.

Investigation 4: *Radiation*   63

Insulation helps keep homes comfortable, no matter what the weather is like outside.

Home Insulation

Brrr! On a chilly winter evening, you might curl up with a blanket, and turn up the heater. Snuggled warm in your home, the cold air outside can't bother you. Houses are designed to help protect the people who live inside.

Heat = Energy Transfer

To heat your home, a number of different energy transfers occur. A heater does its job by transferring thermal energy to the space around it. You don't actually have to touch the heater for it to work. This is because a lot of the heater's thermal energy is transferred by radiation. Radiation can travel to warm you across the room.

The heater also transfers energy to air particles that come into contact with it. This is **conduction**. Then as air warms up, the air begins to move upward. This leads to energy transfer by convection. The heater uses radiation, conduction, and convection to heat the inside of your home.

Now the warm air hits a wall. What happens? When air particles collide with particles in the wall, energy transfers through conduction. Materials inside the walls, windows, floors, and ceiling of a home can reduce energy transfer to the outside wall. These materials are called **insulation**.

Engineering Insulation

Scientists learn about properties of insulation by developing theories and conducting tests. Their evidence and reasoning lead to **models** that explain why certain materials are good insulators. In class, you acted as a scientist when you conducted a controlled test of insulation materials.

You did this by measuring the thermal energy that transferred out of the home system for each material. The better the insulator, the less energy transferred from the home to the outside environment. You analyzed the data to determine which materials provide the best insulation by decreasing energy transfer.

Engineers use scientific knowledge to design solutions to real-life problems. Scientists and engineers often work together to find creative solutions to a problem. In class, you used the engineering design process to come up with a new design to reduce energy transfer.

First you had to understand the **criteria** and **constraints** of the problem. Then you used what you learned about insulation materials and developed a new insulation design. After you collected data from your first test, you looked at real-life homes and thought about how to incorporate a roof into your design. Then, you used this information and the results of your first test to improve your design. You tested that design, too.

Engineers evaluate each proposed solution. How well does it meet the criteria and constraints of the problem? Engineers don't stop until they have developed a working solution.

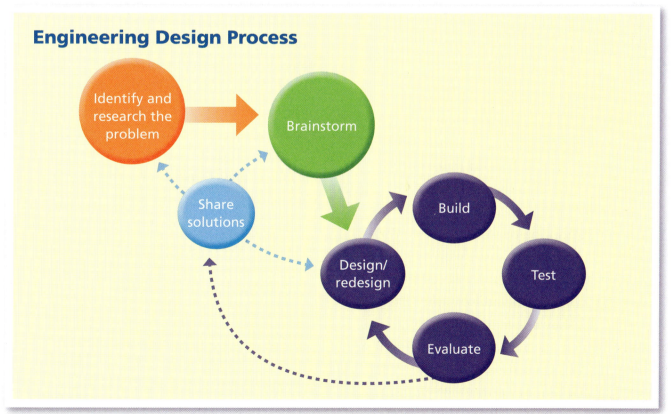

The engineering design process is a series of steps that engineers (and science students!) follow to solve problems and improve solutions.

Investigation 5: *Conduction*

Insulation material in walls, roofs, and attics reduces energy transfer and heating or air conditioning costs. Fiberglass batting is an inexpensive, easily installed insulation material.

Insulation in Buildings

In the average home, half of all the energy used is for heating and cooling. Insulation reduces energy transfer between the inside and outside environments. A well-insulated home stays warm during cold months and cool during hot months. For this reason, well-insulated buildings are more energy efficient than poorly insulated buildings. People in these buildings don't need to use the heater or air conditioner as much, which lowers their energy use.

Energy cannot be created, only transferred, which is why reducing energy use is so important. Some energy sources are **nonrenewable**, which means once they are used up, they are gone. Examples are coal and oil, which are burned to generate heat or electricity. **Renewable** energy sources, like heat from the Sun or Earth, don't run out. The energy source is another factor engineers can include in their design.

Engineers have to consider many factors when designing insulation for an entire building. Criteria include how large the building will be, the local climate, and fire safety. Constraints include the cost of materials and labor, health concerns related to materials, and construction deadlines. Engineers can choose from many good insulating materials, as you learned in class.

Some materials used for building insulation are brand new. Others are recycled, like sugarcane fiber or denim fiber from old jeans. The overall design of a building can also improve energy efficiency. Making the walls thick is one simple technique. But windows are holes in the walls, presenting a challenge for engineers. One solution is to use double-paned windows. Double-paned windows insulate about twice as well as single-paned windows. Why is that?

Design of a Thermos

A thermos will help you understand the benefit of double-paned windows. If you look into a modern thermos, you will see a shiny liner. The outside layer is usually stainless steel or plastic. What you can't see is the space between the two layers. What special material insulates that space? Nothing! When the thermos is made, the air between the two layers is removed. Then the layers are sealed so nothing can get back in. We call this sealed, empty space that contains no particles of air or anything else a vacuum.

Why would a vacuum be the best insulator? A vacuum has no particles to transfer energy between the liquid in the container and the surrounding environment. Energy can still transfer by conduction through the thermos lid and the walls of the container, as particles of liquid transfer energy to the particles of these solid parts. But the vacuum layer slows the rate of overall conduction.

Take Note

Think about conduction, which is energy transfer by contact between particles. Write down your ideas about why a vacuum affects conduction.

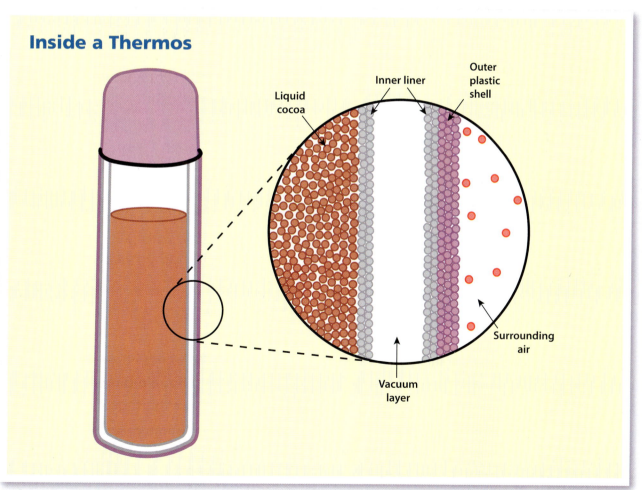

Inside a Thermos

A thermos can keep soup hot or juice cold. How? Inside is a double-walled container. The space between the walls is a vacuum—nothing, no particles. The vacuum reduces energy transfer into or out of the container.

Investigation 5: Conduction   67

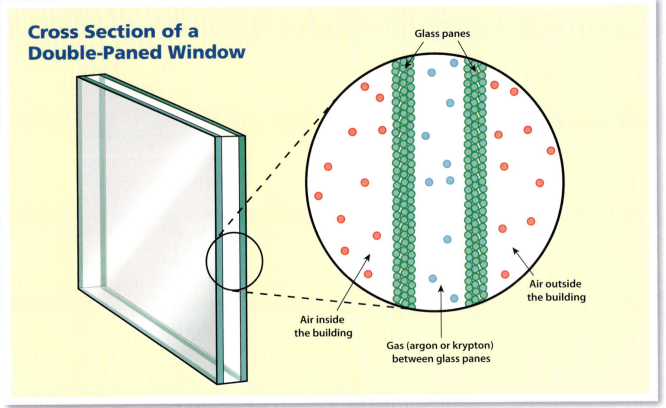

Cross Section of a Double-Paned Window

Glass panes

Air inside the building

Gas (argon or krypton) between glass panes

Air outside the building

Energy transfer through windows accounts for about 25 percent of heating and cooling costs for the average home. The gas layer in a double-paned window cuts down energy transfer by conduction.

Decreasing Conduction

Let's go back to the window. Particles of glass in a single-paned window rapidly conduct kinetic energy from the warm side to the cold side of the pane as particles within the pane collide with each other. Double-paned windows have a space between two layers of glass, like a thermos. The space between panes is not a vacuum, however. Windows need to be strong and watertight. The space between panes is filled with a substance that has very few particles spread out over a lot of space. What substance meets this criterion? (Hint: Think about solids, liquids, and gases.)

The space in most double-paned windows is filled with a very stable gas, like argon (Ar). Some windows are filled with krypton (Kr). Using a gas between two thin panes reduces conduction and provides better insulation than a single, thicker pane. So engineers have developed a good solution: the double-paned window. Does that mean they will stop coming up with new ideas? No! Engineers are always improving designs. They are now designing triple-paned windows for even better insulation. Your class developed an insulating design in just a few days. Imagine what you might find if you kept testing new designs!

Think Questions

1. How does a heater warm a home?
2. How is a double-paned window similar to a thermos?
3. What are some criteria and constraints that engineers must consider when designing home insulation?

Heating the Atmosphere

The campfire has burned down to a bed of hot coals, perfect for toasting some marshmallows.

You get a stick about 1 meter (m) long, to keep your hand away from the hot coals. After a minute, the marshmallow is brown and gooey. You pop it into your mouth. Yikes! You didn't wait long enough for it to cool.

This story has two intense experiences with heat. It shows two ways that energy can move. (1) The coals heated your skin from a distance. (2) The marshmallow transferred energy directly to your tongue when you tried to eat it.

Energy transfers three ways: convection, conduction, and radiation. Which do you think is at work when embers toast a marshmallow? When a freshly toasted marshmallow burns your tongue?

Investigation 6: Air Flow

Energy Transfers

Objects in motion have kinetic energy. The faster something moves, the more kinetic energy it has. Particles are always in motion. They vibrate back and forth in solids and move all over the place in liquids and gases. The amount of kinetic energy in the particles of a material determines its temperature. In hot materials, particles are moving fast. In cold materials, they are moving more slowly.

Conduction. Kinetic energy naturally transfers from a hotter location to a cooler one. In our story, the energy in the hot marshmallow transferred to your tongue. This memorable experience is an example of what happens when energy moves from one place to another by direct contact. That kind of energy transfer is called conduction. The faster-moving particles of the marshmallow transfer energy to the slower-moving particles of your tongue. If the particles of your tongue gain too much kinetic energy, you feel it as pain. At the same time, particles of the marshmallow lose kinetic energy. The marshmallow cools off. Another example of conduction is when you jump into a cold lake and energy moves from your body to the water. Brrrr!

A painfully burned tongue is a one-way transfer of energy, from hot marshmallow to cool mouth. Blowing on food before biting can help, by increasing energy transfer to the surrounding air.

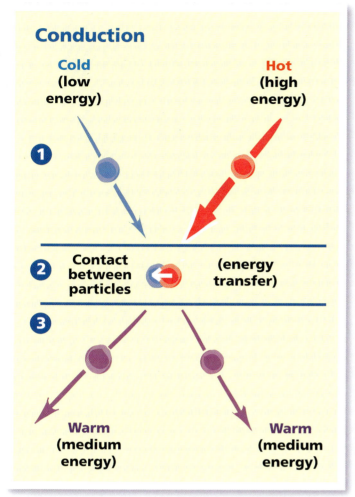

Conduction

A particle in a hot material with a lot of kinetic energy collides with a particle in a cold material with little kinetic energy. Energy transfers at the point of contact. The cold-material particle then has more energy, and the hot-material particle has less energy.

Radiation. What about that energy from the hot coals, which you felt from a distance? That energy traveled right through air without any direct contact. That is called thermal energy, which is a form of **radiant energy**. Radiant energy can travel in rays straight from sources like the hot coals, lightbulbs, and the Sun. Rays from the Sun travel right through the empty space between Earth and the Sun. Some of the Sun's rays are visible light rays. Some are invisible, such as **infrared** rays (beyond the red end of the rainbow) and ultraviolet rays (beyond the violet end of the rainbow).

Heating the Atmosphere

Our atmosphere heats up by these same two kinds of energy transfer: radiation and conduction. Starting with radiation, visible light rays from the Sun pass right through our atmosphere. They strike Earth, get absorbed, and heat up the land and ocean. The heated land and ocean come in contact with air particles. That conduction transfers energy to the air.

The Sun is a source of radiant energy. This energy provides light and heats Earth.

Take Note

What are two ways energy can transfer from one material to another? Can you think of other examples of conduction? Radiation? Record your ideas for each question in your notebook.

The Greenhouse Effect

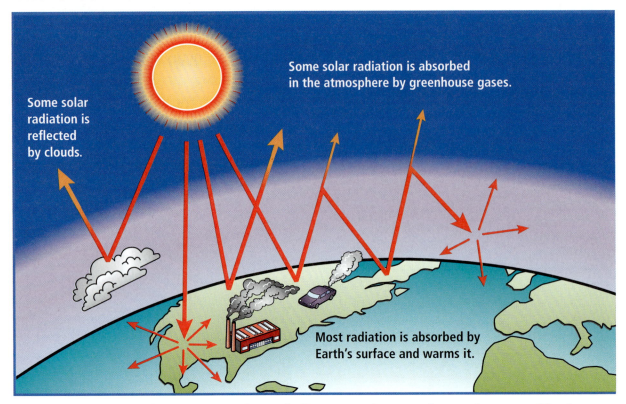

Some of the solar radiation passes through the atmosphere, and some is trapped in the atmosphere by greenhouse gases. The effect of this is to warm Earth's surface and the atmosphere.

Infrared rays from the Sun are a different story. Many infrared rays travel through the atmosphere like visible rays do, and heat the land and the ocean. But some infrared rays don't make it all the way to Earth's surface. They get absorbed by gas particles in the atmosphere. Gases that can absorb infrared rays are called greenhouse gases. Carbon dioxide and water vapor are both greenhouse gases. When they absorb infrared rays, they heat up the atmosphere.

Remember the heated land and ocean? These warm materials radiate thermal energy as infrared rays. This is very important. Greenhouse gases absorb these infrared rays, just as they absorb the infrared rays coming directly from the Sun. The atmosphere heats up more.

Greenhouse Gases

Greenhouse gases have earned this nickname because they "trap" thermal energy in Earth's atmosphere. Here is how that happens. The greenhouse gases in the atmosphere, including water vapor and carbon dioxide, absorb infrared rays (heat up). They reradiate the thermal energy in all directions, including back toward Earth. Say an infrared ray was headed away from Earth, perhaps after being radiated from a hot asphalt schoolyard. It could be absorbed and reradiated instead of leaving the atmosphere. The more greenhouse gas there is, the more likely this is to happen.

Without any greenhouse gases, Earth would lose a lot more thermal energy to space. Earth's average temperature would be about 33 degrees Celsius (°C) colder than it is. Compare this to the Moon. There the temperature ranges from 123°C to −233°C (average −55°C). Because there is no atmosphere, all the thermal energy from the day radiates back into space during the night. By comparison, greenhouse gases keep Earth at a warmer stable temperature that permits life.

Energy Gets Around

So visible light rays and infrared rays heat the surface of Earth through radiation. In addition, energy transfers to greenhouse-gas particles in the atmosphere when they absorb infrared radiation. This radiation is not absorbed by nitrogen and oxygen particles, which make up most of the atmosphere and are not greenhouse gases. But all gases in the atmosphere heat up when they come in contact with the heated surface of Earth. The energy transfers through conduction. In addition, greenhouse-gas particles that absorb infrared radiation can transfer energy to nitrogen and oxygen particles through conduction.

Energy continually transfers between Earth's systems due to interactions among our planet's air, land, and water—as well as living things.

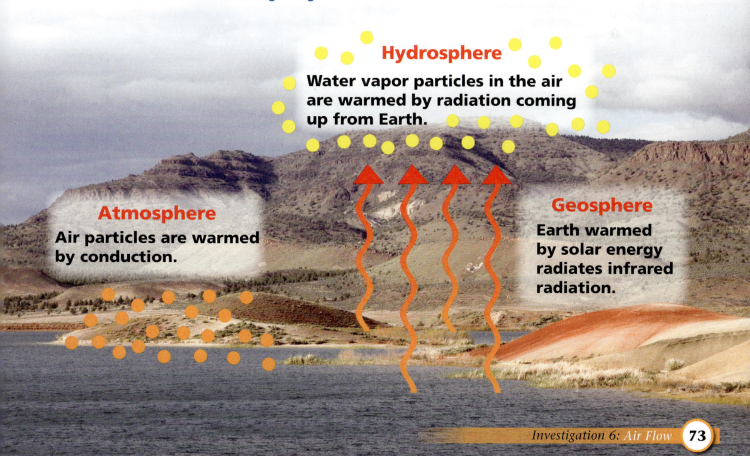

Hydrosphere
Water vapor particles in the air are warmed by radiation coming up from Earth.

Atmosphere
Air particles are warmed by conduction.

Geosphere
Earth warmed by solar energy radiates infrared radiation.

The final piece of the puzzle is convection. Convection moves masses of air or water, both fluids. At the hottest parts of Earth's surface, huge masses of air heat up, expand, get less dense, and move upward. This starts up the largest convection cells on Earth. Over the ocean, large amounts of heated water vapor rise, riding along in the giant convection cell. When the air cools, water vapor condenses into droplets of liquid water, forming clouds. Convection cells redistribute thermal energy and water around the planet.

> **Take Note**
>
> **Explain two ways that Earth's atmosphere gets heated. How do greenhouse gases help keep Earth warm? Record your ideas in your notebook.**

A typical coal-burning power plant releases 3.5 million tons of carbon dioxide into the air every year.

Greenhouse Gases and Climate Change

Greenhouse gases help keep Earth warm. This affects humans and other organisms on Earth. If we change the amount of greenhouse gases, are we changing the amount of energy in the atmosphere?

One greenhouse gas that humans regularly produce is carbon dioxide. It is released every time we turn on a gasoline-powered car or burn coal to generate electricity, among other sources. If you imagine all the electricity and gas used by each of us in industrialized nations every day and multiply that by the number of people . . . well, you can imagine why the amount of carbon dioxide in the atmosphere is increasing. Scientists have tracked carbon dioxide in the atmosphere since the 1950s. The data are shown in the graph.

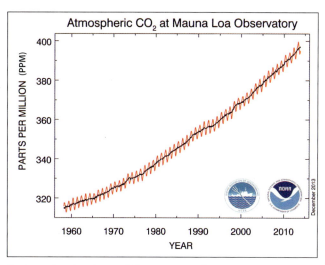

Scripps Institution of Oceanography
NOAA Earth System Research Laboratory

Think Questions

1. What patterns do you see in the data for atmospheric carbon dioxide?
2. What might happen on Earth if humans increase the amount of greenhouse gases in the atmosphere?

About one quarter of greenhouse gas output comes from burning fossil fuels for transportation.

Investigation 6: Air Flow

Wind on Earth

What's the weather going to be like tomorrow? If it was cloudy to the west yesterday, are the clouds likely to head toward you or away? Can prevailing winds help you forecast the next day's weather?

Remember the air-flow pattern that you observed in the convection chamber. A candle at one end of the chamber heated the air. Cool, smoke-filled air entered the chamber at the other end. It immediately sank to the bottom of the chamber. The cool

Wind is a major weather factor and also moves weather systems. Some types of severe weather are classified by wind speed: hurricanes have winds of 119 km per hour or greater.

air moved across the bottom of the chamber and under the warmer, less-dense air. It lifted the warm air. The rising warm air had no place to go, so it traveled across the top of the chamber to the other end. There the air cooled and sank. The whole cycle repeated, producing a small convection cell.

This is a miniature model of what happens in the atmosphere. You have seen that earth materials heat and cool at different rates (remember the experiment with sand, soil, water, and air). You also drew diagrams of local winds to show how this differential heating creates convection cells that cause wind.

Land and ocean in the tropics absorb a lot of solar energy, causing them to heat up. Air above the land and water heats up, then moves upward in the atmosphere.

Global Convection

Now let's look at differential heating on a global scale. Because of the high solar angle, Earth's surface is always warmer in the tropics, the part of the planet near the equator. Tropical landmasses and water masses in the tropical ocean absorb a lot of solar energy. Air in contact with the tropical sea or land receives a lot of energy by radiation and conduction. Huge masses of air heat up and begin to rise. The largest convection cell on Earth starts here. It is the size of the entire equatorial area.

The equatorial convection cells circle the globe like two bicycle inner tubes. Because much of the energy transfer occurs over the ocean, large amounts of water vapor rise high into the atmosphere, riding along in the convection cells.

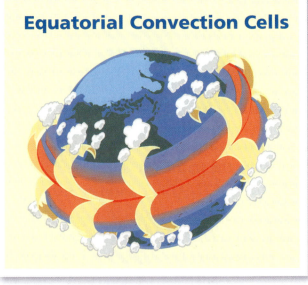

The convection cells that surround the equator are the largest on the planet.

78

As the warm, moist air rises, it cools. At about 10 kilometers (km) altitude, the warm air has cooled to the same temperature as the surrounding air. The cool air begins to fall back to Earth. But because a wall of warm air is rising from the tropics, the cool air cannot fall back directly. The cool air, still at about 10 km, flows north and south, like two gigantic conveyor belts. When it reaches about 30° north and 30° south, it descends toward Earth's surface.

Meanwhile, the warm, low-density, rising air in the tropics creates a low-pressure area. The cooler, more-dense air that is descending from the upper atmosphere creates a high-pressure area. The more-dense air flows into the area of low pressure to replace the rising air. This creates a huge convection cell called a **Hadley cell**, named after George Hadley (1685–1768). He proposed the idea of these enormous convection cells in 1735. The bottom of the cell flows across the surface of the planet, from about 30° north and south toward the equator, producing wind.

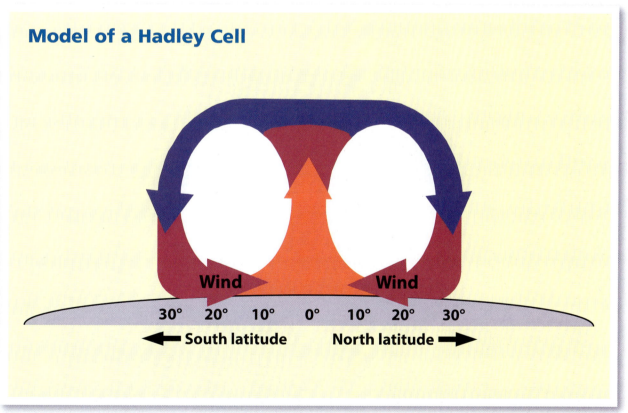

This diagram is another way of looking at the convection cells shown on the previous page. It shows the air flow at the equator from a different perspective, from a position at the surface of Earth. Can you see why winds in the tropics typically blow toward the equator?

Investigation 6: *Air Flow* 79

Prevailing Winds

Differential heating creates high- and low-pressure areas, causing wind. Winds always tend to move from a high-pressure area to a low-pressure area. However, winds do not blow directly from high-pressure areas to low-pressure areas. Instead, the wind curves. Why does this happen?

As air in the atmosphere moves, Earth is rotating, dragging the air along with it and changing its direction. This is called the **Coriolis effect**. This easterly movement of the surface of the planet means the winds in the Hadley cell do not go straight north or south. The Coriolis effect causes the winds in the Northern Hemisphere to curve clockwise and the winds in the Southern Hemisphere to curve counterclockwise.

The combination of high- and low-pressure areas and the Coriolis effect controls the **prevailing wind** direction. Prevailing winds are global winds. They are predictable for different latitudes on Earth. These global winds are not greatly influenced by structures on Earth's surface, like forests, cities, and mountains. Prevailing wind direction is important to people sailing ships, flying balloons around Earth, piloting airplanes, and building wind turbine farms to generate electricity. As you will see later, prevailing winds are also an important factor in the local weather.

Trees can be living indicators of prevailing wind direction.

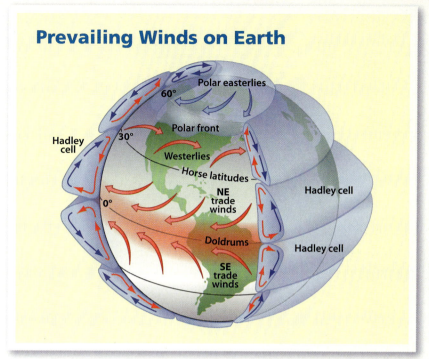

Prevailing Winds on Earth

Prevailing winds impact weather and climate by moving heat and moisture in predictable patterns over Earth's surface.

Winds of the World

Centuries ago, sailors named the different prevailing winds. Between 5° and 25° north or south latitude, dependable winds usually carried a ship across the ocean quickly. Because ships were crossing the ocean carrying goods to trade between the Americas and Europe, these winds became known as the **trade winds**.

Near the equator, the air is rising. At about 30° north and south latitudes, the air is descending. At these latitudes, the wind is usually very light or nonexistent. Sailing ships in these areas could get stuck for days or weeks. The calm area around the equator became known as the **doldrums**. The windless area around 30° north was known as the **horse latitudes**. This term might have been coined when Europeans were carrying horses across the ocean to their colonies. Ships would sometimes get stuck at about 30° north latitude, making the trip much longer than planned. The sailors would run out of water for the horses, and horses could die.

Prevailing winds and Hadley cells happen at the surface of Earth. There are winds at upper levels of the atmosphere as well. The most familiar one is the jet stream. Jet streams are narrow bands of high-altitude wind that occur in predictable patterns, but vary over time. Changes to the jet streams can greatly affect weather patterns on the surface of Earth. They can even change the flight time of aircraft. Flights from the west coast to the east coast typically take less time than the other direction, because they are boosted by the jet stream's air flow.

In addition to the Hadley cells, the Northern and Southern Hemispheres have two other bands of convection cells. They produce prevailing westerlies (wind blowing from the west) between 35° and 55° north and south latitudes and polar easterlies (wind blowing from the east) north and south of 60°.

Chicago's climate is affected by its location near a large body of water.

Local Winds

Local winds change with the season and even with the time of day. They are the direct result of local differential heating. Local winds are affected by land structures and bodies of water near landmasses.

A typical Chicago weather forecast in the summer might go something like this: "Sunny skies today with a high of 29°C inland. Temperatures will be in the low 20s lakeside." Chicago's climate is affected by its location near a large body of water. Landmasses near the Great Lakes or the ocean feel the cooling effects of the nearby water.

From your experiment with heating earth materials, you know that the land gets hotter faster than water when the Sun shines. The hot land transfers energy to the air above it, and the air expands. The warmer, less-dense air creates a low-pressure area over the land.

Air Density Differences

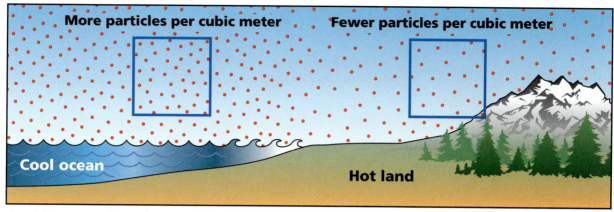

A cubic meter of warm air has fewer particles than a cubic meter of cold air. Warm air is less dense than cool air.

Sea Breeze

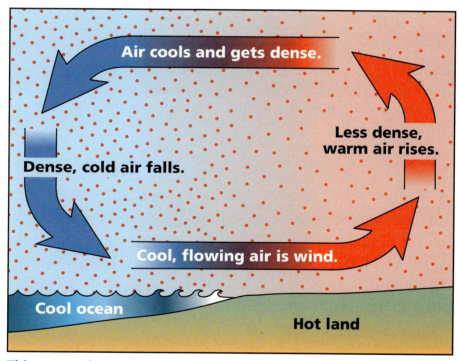

This convection current creates wind that blows from the ocean to the land.

Even though the sea also absorbs energy, its temperature does not change much at all. During the day, the sea stays cooler than the land. Less energy transfers to the air over the water. It remains cool and dense, which means higher air pressure. Air from the high-pressure area over the water flows into the area of low pressure. Wind blows from the water onto the land. This is called a **sea breeze** or offshore breeze.

Land Breeze

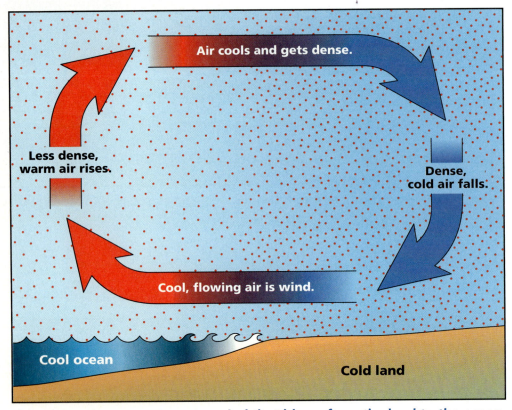

This convection current creates wind that blows from the land to the ocean.

At sunset, there may be a period of calm when land and sea temperatures are about equal. After sunset, the land cools off quickly. The air over the land cools and contracts, becoming more dense. At this point, the local high-pressure area is over the land. Meanwhile, the sea temperature is still about the same as it was during the day, so the air pressure is about the same, but it is lower than the local pressure over the land. Now the local high pressure over the land moves into the lower-pressure area over the water. Wind blows from the land out to sea. This is called a **land breeze** or onshore breeze.

Think Questions

1. What causes the low-pressure region found throughout the tropics?
2. Why is there almost no wind at the equator and at 30° north latitude, but a dependable northeasterly wind in between?
3. What causes local winds to blow from the sea onto the land during the day?

The National Weather Service takes upper air observations with radiosondes suspended from weather balloons. A radiosonde is a sturdy, shoebox-size, cardboard package. Inside is a mini-weather station.

Weather Balloons and the Radiosonde

Weather balloons collect weather data from the troposphere. Twice a day, all over the Earth, weather balloons are released at exactly the same time.

In the late 19th century, meteorologists used kites to gather data about the air above them. Kites could fly up to 3 kilometers (km) into the air. Instruments on the kite gathered data on temperature, air pressure, and humidity. Kites worked well when the wind cooperated.

The **radiosonde** was developed in 1943. A radiosonde is a bundle of weather instruments in a small package. A balloon carries it into the stratosphere. It has sensors for measuring temperature, relative humidity, and air pressure. Measurements are taken continuously as the balloon rises. A radio transmitter sends the data to a ground receiver. A tracking device monitors the location of the radiosonde during its flight. Wind speed and direction are calculated from the tracking data.

A weather balloon is made of a thin membrane of rubber. It is filled with either hydrogen or helium. The radiosonde has a biodegradable plastic parachute. The balloon expands as it rises. When the balloon bursts, the parachute carries the radiosonde to Earth.

Investigation 7: Water in the Air

A radiosonde can be used up to seven times. About one-third of the radiosondes launched by the National Weather Service (NWS) are recovered. Instructions on each radiosonde explain how to return it to the NWS. Then, the NWS gets the radiosonde ready for another flight.

There are more than 900 upper-air observation stations around the world. Most stations are in the Northern Hemisphere. The United States has 108 of them. Observations from these stations are taken at the same times every day of the year, at 00:00 and 12:00 Coordinated Universal Time. The data have many uses, including global and regional weather forecasts, navigation, weather research, and climate-change studies.

New technology allows balloon launchers to track and collect the equipment. Radiosonde technology is still inexpensive and efficient enough to be used for worldwide launches.

Think Questions

1. What is a radiosonde?
2. When do meteorologists launch weather balloons?
3. What are some of the advantages of using balloons to gather weather information?
4. Why do weather balloons expand as they rise through the troposphere?

In 1917, this box kite was used to measure humidity. It could only rise about 3 km and had to be reeled in by its piano-wire string before data could be evaluated.

Animal Rains

Can you imagine fish raining from the sky? How about spiders or frogs?

It may sound like something in a science fiction novel or a fake news story. But reports of animals falling from the sky are well-documented throughout history.

In 1873, *Scientific American* reported that Kansas City, Missouri, was blanketed with frogs that dropped from the sky during a storm. In 1861, fish rained from the sky above Singapore. In 1901, residents described a rainstorm in Minneapolis, Minnesota, that produced frogs to a depth of several inches. In 1948, golfers in England found their golf course covered in herring (a kind of fish) after a light shower. In 1953, Leicester, Massachusetts, was hit with a downpour of frogs and toads of all sorts, which residents claimed choked rain gutters. In 1981, citizens in southern Greece awoke to small green frogs from North Africa, falling from the sky. And more recently, in 2007, hikers described spiders raining from the sky in Salta Province, Argentina.

In a rainstorm, precipitation in the form of liquid water falls from clouds to Earth's surface. In an animal rain, the precipitation hops, swims, and crawls!

Investigation 7: Water in the Air

Most animal rains involve fish and amphibians, like frogs.

Assuming that at least some of these accounts are true, what causes these strange events?

> **Take Note**
>
> Look back in your notebook. Can you think of any weather conditions you have studied that could explain animals falling in rainstorms?

Hypothesis

If you guessed that wind plays a part, you are right! The most likely explanations involve strong winds traveling over water. These winds could pick up small creatures such as fish or frogs and carry them for several kilometers. Remember the videos about tornadoes and hurricanes that you watched earlier? They showed that sometimes wind grows stronger and stronger in a cycle, causing surprising effects. That important concept is a key factor here.

Did you also guess that water was involved? The rain animals typically live in or near water. A wind condition that forms over water is the most likely suspect for these animal rains.

Rare tornado-like waterspouts can arise when winds swirl into a strong **vortex** across a body of water. The vortex could be strong enough to suck up air, water, and small objects like a vacuum. The vortex can carry objects in the air until cooling air forms clouds. Then precipitating water may drop the airlifted objects.

Evidence

Tornadoes have rarely been seen near animal rains. Waterspouts can have winds that are also strong enough to lift small objects and water. People have seen the falling animals, but no reports say that someone saw an updraft. Many questions arise about how the updrafts occur, and how high they go.

The evidence is that most animal rains include small, light, aquatic animals and that most are preceded by a storm. This evidence supports the updraft/waterspout hypothesis. But these explanations do not account for all the events. Some animal rains lack water. Some accounts involve heavy fish or land animals. And most involve only one species, instead of a variety of similar-sized animals. It is also remarkable that some reports say that the animals survive their drop from the sky!

One eyewitness from Texas reported, "My experience was coming out to my car after a short rain shower. I found small, partially frozen fish all over my car and the surrounding parking lot. The school where I worked was a couple of kilometers from a large lake. It is believed that a waterspout sucked up a school of the small herring-like fish that feed near the surface. The waterspout had carried them high enough in the atmosphere to partially freeze them. As the waterspout moved over land, it dissipated and the fish fell to the ground."

A waterspout, a type of tornado that forms over water, is powerful enough to pick up and carry fish and frogs. When the funnel eventually loses energy and stops spinning, the "passengers" are released in unexpected places.

Investigation 7: Water in the Air

Other Explanations

Witnesses can be confused about what they saw. Some skeptics believe that if the rain is heavy enough, frogs might look like they are falling from the sky even though they are just jumping around in the rain. In an intense storm, it could be hard to tell the difference. Indeed, some animal rains have been explained in other ways. For example, unusual numbers of spiders could have used their silk as parachutes to travel. And sometimes thousands of worms or frogs emerge in unexpected areas.

Still, these alternative explanations do not account for fish found far from any body of water, or the frozen frogs that pelted Dubuque, Iowa, in 1882. The weather is the likely explanation.

Will you see an animal rain in your lifetime? With so many people carrying cell-phone cameras, maybe you will see an animal rain caught on video. Maybe you will be lucky enough to shoot the video yourself!

Think Questions

1. What is one hypothesis about how animal rains occur?
2. Why do some people think that animal-rain witnesses may be confused about what they saw?

Fish stranded on the ground far from water may be evidence of an animal shower. Depending on how high and how far the animals are carried, they could even fall alive.

This familiar photograph of Earth is the only one ever taken by a human being. It is commonly known as the blue marble image.

Earth: The Water Planet

Earth is known as the water planet. Of the eight planets in the solar system, Earth is the only one that has a vast ocean of water.

Apollo 17 astronauts took this image in 1972 as they left Earth's orbit on their way to the Moon. If this was your first look at Earth, you might think Earth is completely covered with water. Of all the planets in the solar system, Earth is the only one that has a vast ocean of water.

Earth is a nearly closed system. Almost no material comes to Earth from space, and no material is lost. This includes the water. Almost all the water that is here now has been here for billions of years. The good news is that Earth's water is going to stay here for the next several billion years. The bad news is that Earth is not getting any more water from some outside source. What you see is what we've got.

Investigation 8: *The Water Planet*

Where Is Earth's Water?

By now, you know water is almost everywhere. It is in the ocean, in and on the land, and in the atmosphere. The pie charts show how water is distributed on Earth. A quick glance shows that just about all Earth's water is in the ocean. All that water, but we cannot drink it. For us, sea water is toxic.

How do we get the water we need for survival and to support civilization? We need access to the small portion of Earth's water that is fresh. After subtracting the amount of fresh water in ice caps and glaciers and in groundwater, that remaining portion is less than 1 percent of the total water. That small percentage is still a large volume of water, about 13 million cubic kilometers (km^3). This water is in lakes, rivers, swamps, soil, snow, clouds, water vapor, and organisms. It is known as free water because it is free-moving and is being constantly refreshed and recycled.

Most of the water that we can easily use comes from rivers and lakes. Water in rivers and lakes is known as surface water. Water that falls as precipitation can either remain as surface water or seep into the ground. Underground water is known as groundwater. It is stored in soil or porous rock. The pie charts show that there is much more water stored underground than at the surface. It is close at hand, but invisible.

Did You Know?

Humans have been digging wells to gain access to groundwater for at least 8,000 years.

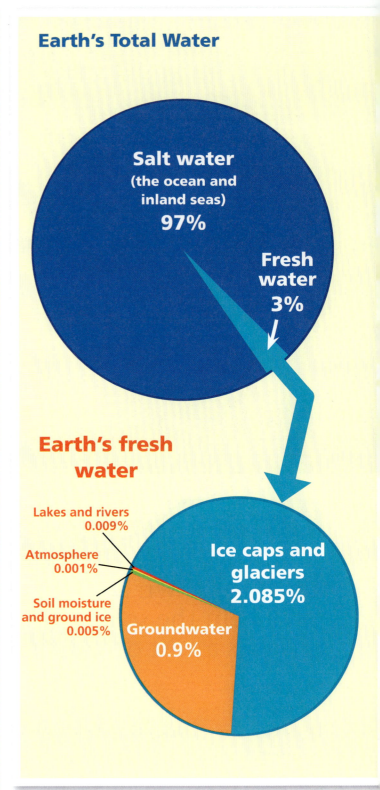

Almost all the water on Earth is either too salty for drinking and watering plants or frozen and unavailable. So we rely on surface water and groundwater to meet our fresh water needs.

Water Use

Americans use a lot of water. Think about this. In 2005, the United States used a total of about 1,550 billion liters (L) of water a day. Over the course of a year, that adds up to more than 500,000 billion L! That translates into half a million cubic kilometers per year. That is a significant 4 percent of all the free water available on Earth.

People use water in many ways. Most important, water is essential for life. Without water to drink, we would not survive. And you can probably think of many nonessential ways you use water at home. You wash clothes, brush your teeth, and cook food with water. Swimming pools are filled with water, and lawns need water to grow. Humans also use water for transportation, for creating electricity, in manufacturing, and for agriculture. Many of these activities require good water quality. And, unfortunately, many of these activities create pollutants that can lower water quality.

> **Did You Know?**
>
> Plastic water bottles are convenient, but they have a cost. Manufacturing and filling bottles uses up to three times as much water as they hold.

People depend on water for life. In fact, 60 percent of the human body is water that must constantly be replenished. Clean drinking water is one of our most valuable natural resources.

Water Distribution

Weather distributes water on Earth's surface. If weather did not continually supply water in the form of rain and snow, all land would be arid and lifeless. Weather does not, however, distribute water equally around the planet. Some places, like the Midwest, have variable water supplies, getting very little precipitation one year and a lot the next. Droughts can cause severe water shortages. The deserts of the world are always dry, while the tropical rain forests are continually soaked. Adding to the unevenness of water distribution is the distribution of human populations. Some densely populated areas, like Los Angeles, California; Phoenix, Arizona; and New York, New York, need more water than is available locally. They have to import water from distant places.

Scientists are concerned about the warming trend on Earth. **Global warming**, due to climate change, could affect both evaporation and precipitation in the United States. If more evaporation than precipitation occurs, the land will dry, lake levels will drop, and river flows will decrease. Other regions may receive more precipitation than usual, creating floods and affecting vegetation. It will take worldwide planning and cooperation to adjust to the effects of climate change.

A flood occurs when water covers land that is normally dry, as when rivers overflow their banks due to heavy rain or snowmelt. Floodwaters enrich the soil but can also destroy crops.

Water Conservation

Earth is the water planet. Fortunately, water is one of our renewable resources. It is constantly being recycled among the atmosphere, land, and ocean. Humans cannot change how much water there is. But we can make smart decisions about how much of it we remove from natural systems, how it is distributed, how it is used, and what happens to it after we use it. The demand for water is increasing worldwide because of population increase and a rising standard of living (which requires water). So everyone will need to participate in water conservation. Industries will need to use water more efficiently. They also need to keep pollutants out of water sources. Agriculture will need to use more conservative crop-watering practices. And every citizen will need to be aware of the value of water and to treat it as the most precious substance on Earth.

Think Questions

1. Earth's human population was over 7 billion in 2016. The United Nations estimates it will reach 9.7 billion by 2050 and 11.2 billion by 2100. What effect might that increase have on freshwater resources?
2. If humans use more fresh water from lakes and groundwater sources, how might that impact the rest of the water cycle? Earth's systems? Other organisms living on Earth?
3. As Earth's human population continues to grow, what are some strategies that might help conserve our fresh water?

Engineering advances help farmers irrigate their crops using water-efficient systems.

Ocean Currents and Gyres

The ocean covers a whopping 70 percent of Earth's surface. As the ocean heats or cools, how does all that water move? And how do those movements affect weather and climate?

The Driving Forces of Ocean Currents

Just as there are global wind patterns, there are global water patterns, called **ocean currents**. Three main forces affect ocean currents: winds, differences in water density, and tides.

Winds. Winds drive currents in the upper 20 percent of the ocean. In coastal areas, local winds can drive currents. In the open ocean, global winds drive currents.

Differences in water density. Deep currents flow in the lower 80 percent of the ocean. Differences in density of water are the biggest force behind deep ocean currents. You learned earlier that convection currents form when fluids have different densities. Similar factors cause ocean water to have different densities.

As you learned earlier, water heats and cools slowly. But even these slow changes will cause temperature differences from one area of the ocean to another. As the Sun's radiant energy warms the ocean, differences in temperature will cause differences in density. This causes convection currents.

Which is more dense, a solution with lots of salt or little salt? Solutions with more salt are more dense. Variations in the concentration of salt, called **salinity**, cause differences in density of ocean water. This also causes currents.

The density differences due to temperature and salinity result in an upward and downward flow of water masses. These currents occur at both deep and shallow ocean levels and are much slower than surface currents.

The effect of temperature and density on currents is greatest in Earth's polar regions. As polar sea ice forms, the surrounding sea water gets saltier. Pure water freezes and separates from the salt, leaving all that dissolved salt in the ocean. Water near the polar regions is more dense not only because it is cold, but because it is saltier. As this cold, salty water sinks, the surrounding surface water moves in to replace it. Once the water sinks, it travels through the ocean basins. It eventually comes back to the surface, about 1,500 years later! This motion drives the deep-ocean currents around the planet in a huge convection current. We call this system the Great Ocean Conveyer Belt.

The ocean is constantly in motion due to wind, tides, and temperature. Steady wind in the same direction produces surface currents. Wind also causes waves, like the one battering this lighthouse.

Investigation 8: *The Water Planet*

Tides. The third main force that affects ocean currents is tides. The gravitational force of the Moon and Sun produce the tides on Earth. The rise and fall of water interacts with the shore to create surface tidal currents. Tidal currents follow regular patterns, flooding when the tide moves toward the land and ebbing when it moves back out to sea.

The shape of the shoreline strongly affects tidal currents. The Bay of Fundy, between Nova Scotia and New Brunswick, has the right shape to produce tides of about 15 meters (m), the largest in the world. As you can imagine, organisms that live in coastal areas are greatly affected by tidal currents.

The movement of ocean water is complicated. The tides, density changes, winds, and the Coriolis effect (caused by the rotation of Earth) explain most of the ocean's currents. What is the final piece? The ocean is broken up by giant landmasses. Just as water in a stream will flow differently if you drop in a large rock, the movement of ocean water is affected by the continents and islands of Earth.

The flow of ocean water is affected by the landmasses it encounters.

Ocean gyres are spiraling circulations of water created by large oceanic currents. Small, temporary whirlpools, like the one shown here, are called eddies.

Local Currents

The patterns of ocean currents affect how fish and other sea life move around the ocean and control how ships travel. Currents sometimes carry swimmers far from the beach. When surface currents run close to shore, a visitor to the beach could experience a **rip current**. These local currents form around low spots, breaks in sandbars, or structures such as piers. Rip currents can move faster than even an Olympic swimmer. If you are caught in one, do not fight it. Rip currents are usually not more than 25 m wide and break up not far from shore, so it is best to swim around them.

Global Current Patterns

Ocean currents can form swirling areas called **gyres**. A gyre (rhymes with tire) is a large system of rotating ocean currents. It is driven largely by winds. There are five main gyres in Earth's ocean.

What causes these gigantic whirlpools in the ocean? Do you notice any patterns in the map of gyres on the next page? All the major gyres occur in the middle of the spaces between continents. As ocean currents hit shallower areas of coast, they turn in a different direction. Think of water sweeping along the coastline in a large-scale current, called a **boundary current**. As the water approaches a continent and changes direction, you get the first part of a swirl that will ultimately form a gyre.

A boundary current of the South Pacific Ocean is the Humboldt Current (or Peru Current). It flows northward along the coast of Chile. As it turns away from Peru, deeper water upwells to replace the westward-moving water. This big upwelling system brings up cold, nutrient-rich deep ocean water. It feeds an abundance of sea life and the largest fishery on Earth. About 20 percent of Earth's fish catch is made possible by the Humboldt Current! Because the current is cold, it cools the weather and climate of Chile and Peru.

A well-known boundary current is the **Gulf Stream** in the North Atlantic. It flows northward along the east coast of the United States and then across the North Atlantic Ocean toward Europe. The warm water of the Gulf Stream affects the climate of the coast from Florida to Newfoundland and the west coast of Europe. It makes the climate much milder than it would otherwise be.

Take Note

How do you think ocean currents affect your local climate? Record your answers in your notebook.

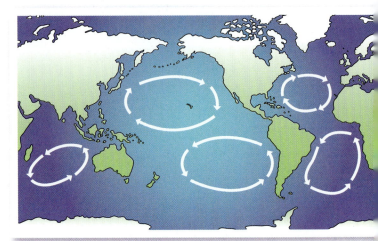

The world ocean contains five huge gyres. They swirl in opposite directions in the Northern and Southern Hemispheres.

Earth's Main Boundary Currents

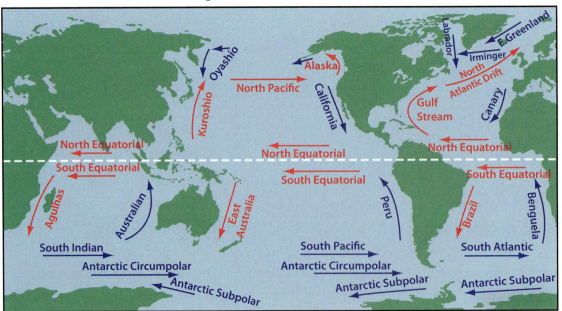

Warm (red) and cold (blue) boundary currents flow along continental coastlines. They affect the temperature of the air above them, which in turn affects local weather and global climate.

Great Pacific Garbage Patches

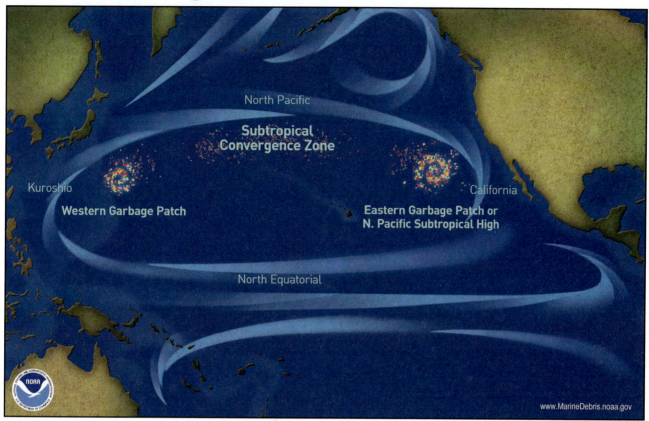

Discarded plastics are carried around the Pacific by four main currents. Eventually, the trash is drawn in by the swirling vortexes of the Eastern and Western Garbage Patches.

Great Pacific Garbage Patches

We see it every day. Someone carelessly drops a plastic cup on the ground as they walk, or tosses a wrapper out their car window. Where are those plastic objects now? Plastic does not biodegrade (break down in nature) quickly. Trash on the ground can get washed by rainwater into storm drains that eventually lead to the ocean.

Imagine all that plastic trash floating into the ocean and traveling far from its source. Where will it end up?

Within the North Pacific Ocean Gyre there are several areas of debris known as garbage patches. These huge swirling currents hold garbage, mostly plastic. The pieces can be as large as huge fishing nets. The vast majority are the size of corn flakes or rice grains, broken from larger plastic objects. The plastic pieces float in the top layer of ocean water and are so tiny that they are nearly impossible to collect.

Ocean litter is not only ugly to look at, it is dangerous to marine animals. Waterborne debris can easily be mistaken for food, resulting in choking, starvation, and other injuries.

Plastic does not degrade quickly. It breaks apart into tinier and tinier pieces, just the right size for sea animals to eat. Fish mistake the tiny plastic particles for food. Toxic chemicals from the plastic leach into fish tissue. Birds such as albatross forage in the North Pacific and eat large amounts of plastic trash, including fishing line, light sticks, and bottle caps. Some birds can throw up the plastic, but thousands die each year from eating plastic garbage. Sea turtles, including rare species at risk of becoming extinct, may think floating plastic bags are jellyfish. Eating the bags can kill the turtles.

> ### Think Questions
> 1. What are three forces that drive ocean currents?
> 2. What are the two main factors that affect deep ocean currents?
> 3. Look at the map of Earth's boundary currents. What would be the best direction to start out for sailors making these trips?
>   - Spain to Florida
>   - China to California
>   - Australia to Africa
> 4. Have you ever seen someone tossing plastic on the ground? What is a possible solution to prevent people from littering?

When ocean birds die and decompose, large amounts of plastic are sometimes found in their stomach contents.

El Niño

Ocean currents have a big effect on weather and climate. The interplay between ocean currents, weather, and climate can be quite dramatic.

The interactions can change weather all over the world. Sudden changes in ocean currents dramatically affect local weather during an **El Niño** year. Usually the trade winds blow strongly from east to west near the equator in the Pacific. The winds pile up water about half a meter high in the western Pacific. At the same time, cold deep water in the eastern Pacific flows upward. It replaces the warmer surface water that is being blown to the west. Normally, the western Pacific has surface water about 8 degrees Celsius (°C) warmer than the eastern Pacific surface water, which is about 22°C.

Every few years, this balance changes because of El Niño. El Niño starts when the equatorial winds weaken. Some of the warm water piled up in the western Pacific starts to slip back east along the equator. As a result, less cold water wells up in the eastern Pacific. The ocean surface temperature is warmer than usual in the eastern Pacific. These warmer surface temperatures are typical during an El Niño year.

Water flow changes in the Pacific can affect climate around the globe.

Effects of El Niño

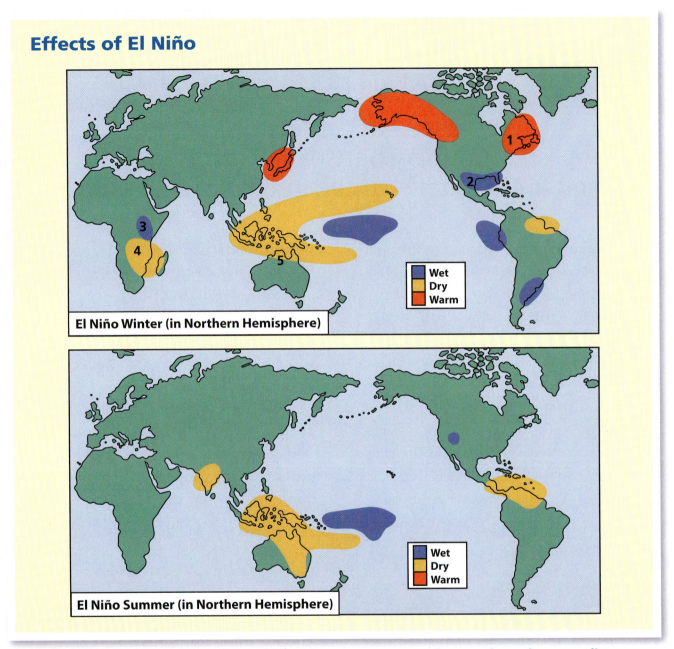

El Niño Winter (in Northern Hemisphere)

El Niño Summer (in Northern Hemisphere)

El Niño influences US weather more in winter than in summer. Is the weather where you live potentially affected by El Niño?

El Niño has dramatic effects worldwide. Peru and Ecuador have heavier-than-normal rains. Less upwelling of cold, nutrient-rich water reduces fish populations. In the northern United States (1 on the upper map), winters are warmer and drier than average. Less snow falls. Winters are cooler and wetter than average in northeast Mexico (2) and the southeast United States. East Africa (3) is wetter than normal from March to May. But south-central Africa (4) is drier than normal from December to February. Drier conditions occur in parts of Southeast Asia and central Australia (5). That leads to more wildfires, worse haze, and poor air quality. El Niño can cause a wetter, cloudier winter in northern Europe and a milder, drier winter in the Mediterranean region.

Climates: Past, Present, and Future

Ocean currents have a big effect on weather and climate. The interplay between ocean currents, weather, and climate can be quite dramatic.

Climate change is in the news and generates a lot of questions. Is it real? Is it natural or caused by humans? Is it happening now? What will happen if we don't stop it? What can we do to stop it?

Climate change is real. The evidence is temperature data over time. The average temperature of the troposphere has gone up by about 0.6 degrees Celsius (°C) over the past 100 years. This may not seem like much, but some weather patterns can be affected by even slight temperature changes. The temperature of the top 300 meters (m) of the ocean has increased by more than 0.3°C in the past 40 years. Even a small increase in global ocean temperature represents an exceptionally large amount of energy.

Climate change can cause glaciers to melt more quickly than they otherwise would.

A big concern is that global warming can increase the rate of melting of ice sheets and glaciers in polar regions. This extra water raises the level of the ocean. Farms, villages, towns, and cities built at or near sea level could be destroyed by the rising water.

Is climate change caused by humans? Although natural factors can cause global warming or cooling, humans contribute, too. The most significant factor in the current warming trend is a human-caused increase in greenhouse gases, which trap thermal energy in the atmosphere. We are increasing greenhouse gases in two important ways. (1) Burning fossil fuels such as coal, oil, natural gas, and gasoline produces carbon dioxide. (2) Trees use photosynthesis to remove carbon dioxide from the atmosphere. They hold it in the form of cellulose for a very long time. Cutting down forests increases the carbon dioxide in the atmosphere.

The South Cascade Glacier in Washington state is one of four benchmark glaciers studied by the US Geological Society. Data collected over decades shows it is retreating rapidly.

Frozen permafrost makes up 24 percent of Northern Hemisphere land! As permafrost melts in a warming world, methane is released in the air, ecosystems are upset, and erosion reshapes Earth's surface.

Earth Systems

The answers to the other questions are a lot more complicated. Scientists are watching Earth systems closely to see how fast climate is changing. Here are some observations.

Ice. Melting ice is a predictable result of global warming. Large chunks of ice are breaking off the Antarctic ice sheets. The North Pole has less ice than ever observed before. Many glaciers (large sheets of ice) are getting smaller. At the present rate of melting, glaciers in Montana's Glacier National Park will be gone in 50 years. Melting also affects permafrost, which is soil that stays frozen, so it is solid and stable. When permafrost begins to melt, there are big problems for people and other organisms who live on it. Melting permafrost releases large quantities of methane, a greenhouse gas.

Sea level. Over the last 100 years, sea level has risen 15–21 centimeters (cm). The increase comes from melting surface ice and from being warmer, which makes water expand. When sea level rises, tides cover more land, and the ocean permanently submerges coastal areas. A large percentage of humans live in large cities along the coasts. The ocean is more likely to flood those cities as sea levels rise. Island dwellers are in particular danger. Also, much of the world's fertile cropland is in low-lying areas.

Permafrost has started to melt and crumble along this shoreline.

> **Try This at Home!**
>
> **Visit FOSSweb for a link to a website that lets you change sea level and witness the effects.**

Ecosystems. As ice and permafrost melt and the ocean warms slightly, all organisms in coastal or permafrost areas are affected. Some species may not have suitable habitats to live in anymore. Others may try to relocate and end up impacting other ecosystems nearby.

What is the difference between global warming and climate change? Global warming is simply an increase in the average temperature of Earth. Climate change is a result of that. Droughts, increased frequency and intensity of storms, hurricanes, and flooding are all predicted changes. Another is changes in the start and length of seasons. These changes affect plant and animal life cycles.

How can humans reduce carbon dioxide emissions?

What's in the Future for Earth?

We can be fairly certain that Earth's temperature will continue to rise as long as the concentration of greenhouse gases rises. You have seen the relationship between carbon dioxide and global temperature in a graph. As we add greenhouse gases like carbon dioxide to the atmosphere, we can expect a continuously warming global climate. We can affect that future only if we stop adding greenhouse gases to the atmosphere and create programs to restore forests that have been burned or stripped away.

The graph here shows atmospheric carbon dioxide over time. It predicts future values. Reducing carbon dioxide emissions could result in the lower-emissions (550 ppm) scenario by the year 2100. Without significant changes in human behavior, the higher-emissions (900 ppm) scenario is more likely. Will we be at the lower-emissions mark or the higher-emissions mark by 2100? That is up to us.

 Go to FOSSweb.com to explore more information about this data.

Carbon Dioxide Concentration for the Last 800,000 Years
(in parts per million, measured from trapped bubbles of air in an Antarctic ice core)

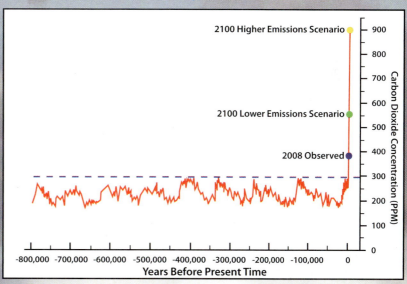

The concentration of carbon dioxide in the atmosphere, stable for millennia, is being multiplied by human activity. It will take global efforts just to slow the upward trend.

Investigation 9: Climate over Time

Change is not easy, but it is possible. Individuals can help by conserving energy to reduce greenhouse-gas emissions. Carbon dioxide is released whenever someone starts a gasoline engine or uses electricity generated in power plants that burn coal, gas, or oil. You are very likely generating greenhouse gases whenever you use the air conditioner, turn on a light, play a video game, charge your cell phone, or travel by car. As individuals, we can conserve electricity, use public transportation, and produce some of our own food in a garden. As a society, we can develop new technologies that do not require fossil fuels that emit carbon dioxide. They might include harnessing more solar-, wind-, and water-powered energy. We can provide better public transportation and make products more efficient.

Think Questions

1. **What do you do that contributes to carbon dioxide emissions?**
2. **What could you do to reduce your carbon dioxide emissions?**

Solar panels and wind turbines capture energy from renewable sources and are alternatives to the burning of fossil fuels. A key advantage of Sun and wind power is that they do not emit greenhouse gases.

Images and Data

Images and Data Table of Contents

Investigation 4: Radiation
World Map 115
Minneapolis-Area Climate 116
Miami-Area Climate 117
US Map with City Locations 118
Great Plains Cities: Climate Data over
    30 Years. 120
West Coast Cities: Climate Data over
    30 Years. 121

Investigation 6: Air Flow
Radar Images of Cloud Cover 122

Investigation 7: Water in the Air
Raindrops and Cloud Droplets 123
Observing Clouds. 124

Investigation 8: The Water Planet
US Map with Western Cities 126
Alaska Climate over 30 Years. 127
Southern California Climate over
    30 Years. 128
Northern California Climate over
    30 Years. 129

Investigation 9: Climate over Time
Climate Graph A 130
Climate Graph B 131

Investigation 10: Meteorology
Fronts 132
Weather and Fronts 133

References
Science Practices. 134
Engineering Practices. 135
Engineering Design Process 136
Science Safety Rules 137
Glossary 138
Index 142

World Map

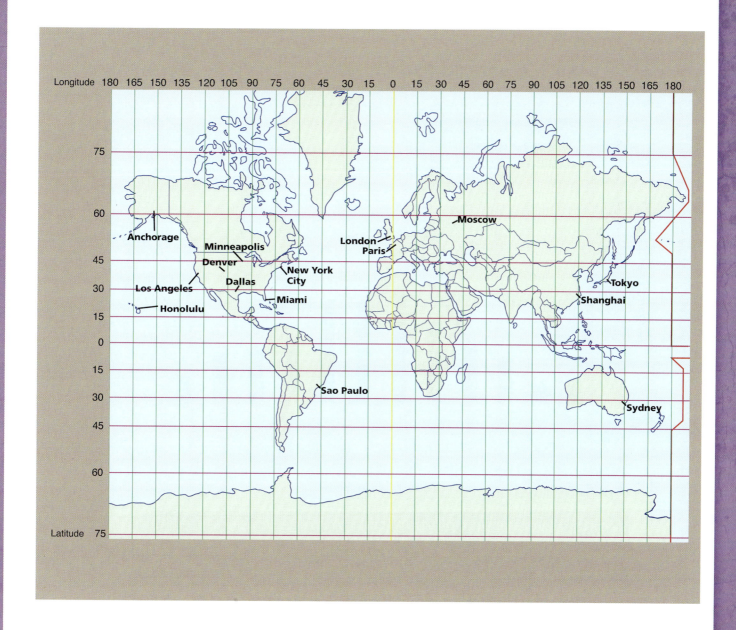

Investigation 4: Radiation

Minneapolis-Area Climate

Climate Graph and Data for 30 Years for Minneapolis, Minnesota

Latitude: 45.0° N   Longitude: 93.3° W

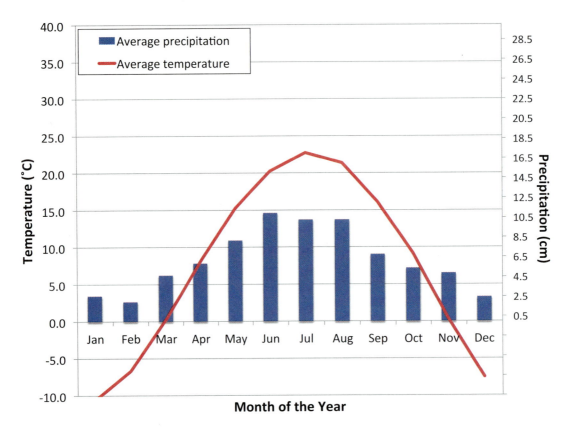

Month	Average precipitation (cm)	Average temperature (°C)
Jan	2.6	-10.6
Feb	2.0	-6.7
Mar	4.7	0.3
Apr	5.9	8.1
May	8.2	15.3
Jun	11.0	20.3
Jul	10.3	22.8
Aug	10.3	21.4
Sep	6.8	16.1
Oct	5.4	9.2
Nov	4.9	0.3
Dec	2.5	-7.5

Miami-Area Climate

Climate Graph and Data for 30 Years for Miami, Florida

Latitude: 25.8° N   Longitude: 80.2° W

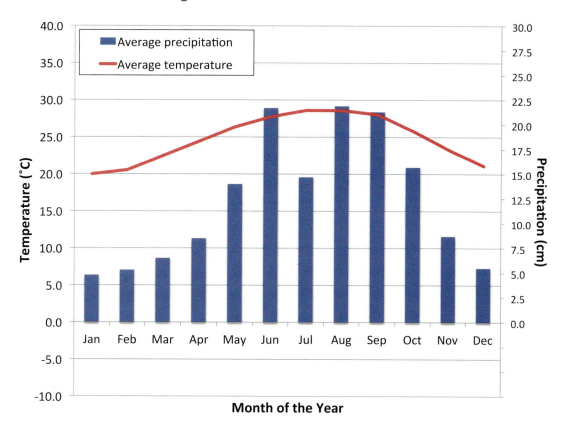

Month	Average precipitation (cm)	Average temperature (°C)
Jan	4.8	20.0
Feb	5.3	20.6
Mar	6.5	22.5
Apr	8.5	24.4
May	14.0	26.4
Jun	21.7	27.8
Jul	14.7	28.6
Aug	21.9	28.6
Sep	21.3	28.1
Oct	15.7	25.8
Nov	8.7	23.3
Dec	5.5	21.1

Investigation 4: Radiation   117

US Map with City Locations

Great Plains Cities: Climate Data over 30 Years

Sioux Falls, South Dakota

Latitude	43.5° N
Longitude	96.7° W
Elevation	430 m

Average temperature in °C

Jan	Feb	Mar	Apr	May	Jun
-10.0	-6.1	0.3	7.8	14.4	20.0

Jul	Aug	Sep	Oct	Nov	Dec
22.8	21.4	16.1	8.9	2.5	-7.5

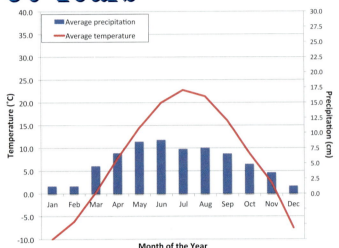

Salina, Kansas

Latitude	38.8° N
Longitude	97.6° W
Elevation	370 m

Average temperature in °C

Jan	Feb	Mar	Apr	May	Jun
-1.7	1.7	7.2	12.8	18.1	24.2

Jul	Aug	Sep	Oct	Nov	Dec
27.2	26.4	21.1	14.4	6.4	0.3

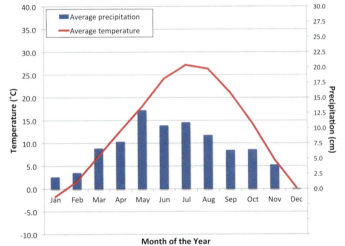

Dallas, Texas

Latitude	32.8° N
Longitude	96.8° W
Elevation	160 m

Average temperature in °C

Jan	Feb	Mar	Apr	May	Jun
7.5	10.6	15.0	18.9	23.6	28.1

Jul	Aug	Sep	Oct	Nov	Dec
30.3	30.0	26.1	20.3	13.6	8.9

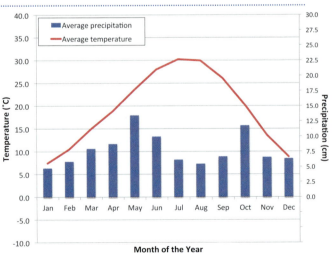

West Coast Cities: Climate Data over 30 Years

Newport, Oregon

Latitude	44.6° N
Longitude	124.1° W
Elevation	3 m

Average temperature in °C

Jan	Feb	Mar	Apr	May	Jun
7.2	7.8	8.6	9.4	11.4	13.1

Jul	Aug	Sep	Oct	Nov	Dec
14.4	14.7	13.9	11.7	9.2	7.2

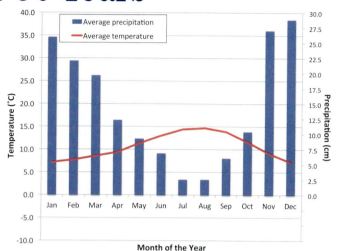

San Francisco, California

Latitude	37.8° N
Longitude	122.4° W
Elevation	3 m

Average temperature in °C

Jan	Feb	Mar	Apr	May	Jun
11.1	12.8	13.1	14.2	14.4	15.8

Jul	Aug	Sep	Oct	Nov	Dec
16.1	16.9	17.5	16.9	14.2	11.7

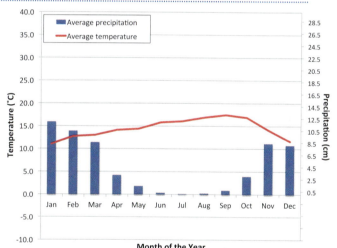

San Diego, California

Latitude	32.7° N
Longitude	117.2° W
Elevation	3 m

Average temperature in °C

Jan	Feb	Mar	Apr	May	Jun
13.1	13.3	13.9	15.3	16.9	19.2

Jul	Aug	Sep	Oct	Nov	Dec
21.7	22.2	21.7	18.9	15.6	13.1

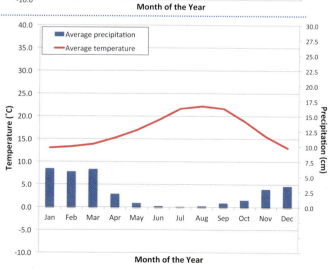

Investigation 4: Radiation

Radar Images of Cloud Cover

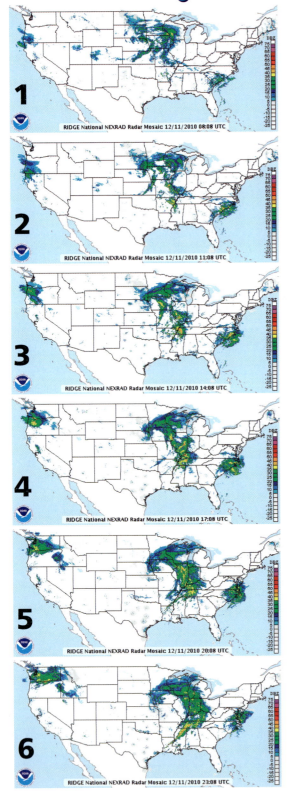
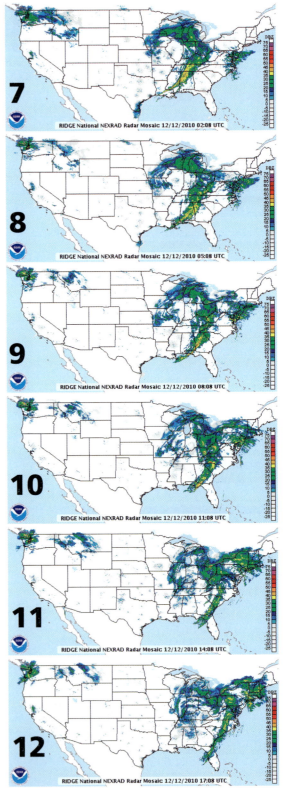

Raindrops and Cloud Droplets

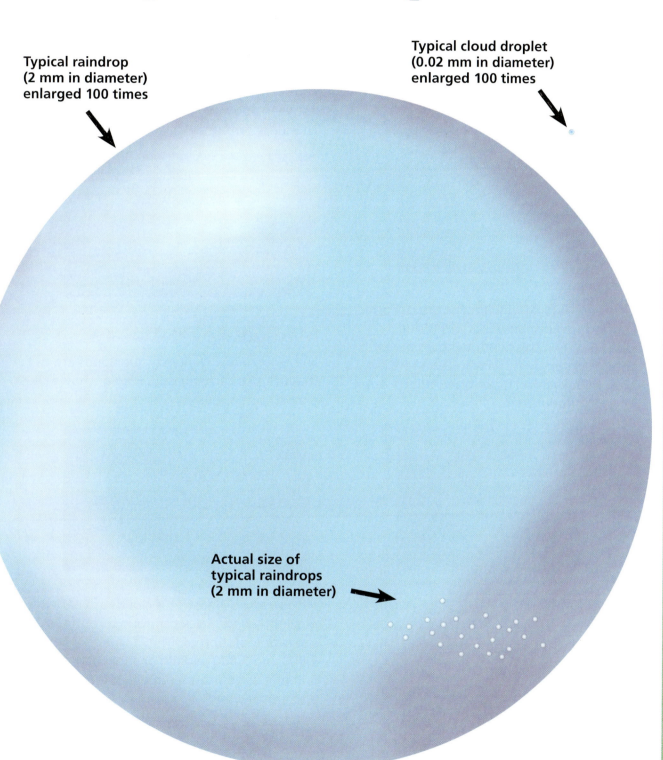

Typical raindrop (2 mm in diameter) enlarged 100 times

Typical cloud droplet (0.02 mm in diameter) enlarged 100 times

Actual size of typical raindrops (2 mm in diameter)

Investigation 7: *Water in the Air*

Observing Clouds

Low-Level Clouds

Stratus. The base of stratus clouds is often around 600 m. Stratus clouds form in stable air. They appear flat and layered, with no lumps or bumps.

Stratocumulus. Stratocumulus clouds form when warm, moist air mixes with drier, cooler air. When this mixture moves beneath warmer, lighter air, it starts to form rolls or waves. It looks thick. It can bring drizzle or light precipitation.

Cumulus. Puffy white clouds at low levels are called cumulus clouds. When they are small and scattered, it means good weather. These are sometimes called fair-weather cumulus.

Cumulonimbus. Cumulonimbus clouds form on hot summer days with little wind. Air heated by the ground rises. Convection cells form. Warm air rises through the cell center. Cooler air sinks down the sides. It has a base between 300 and 1,500 m. Rain starts to fall. Thunderstorms may develop.

Medium-Level Clouds

Nimbostratus. Nimbostratus clouds are sheet clouds carrying rain. Rain or snow falls almost continuously.

Altostratus. Altostratus clouds appear white or slightly blue. They can form a continuous sheet or look fibrous. They form between 2,000 and 5,000 m. Rain or snow may fall. Sometimes you can see the Sun through an altostratus cloud.

Altocumulus. Altocumulus clouds form between 2,500 and 5,500 m. They look like small, loose cotton balls floating across the sky.

Altocumulus mammatus. Altocumulus mammatus clouds look threatening, but actually indicate that the rainy weather is almost over. The clouds droop because the air is cooling and sinking.

Altocumulus lenticularis. Altocumulus lenticularis clouds are lens-shaped. Sometimes they look like flying saucers. They form at the top of a wave of air flowing over a mountain peak or ridge.

High-Level Clouds

Cirrus. Cirrus clouds are made of falling ice crystals. The wind blows them into fine strands. The longer the strands, the stronger the wind. Cirrus clouds indicate that the air is dry. Good weather should continue.

Cirrocumulus. Cirrocumulus clouds are high-level clouds composed of many smaller clouds. Some people think these clouds look like fish scales. A sky full of cirrocumulus clouds is sometimes known as a mackerel sky. It might mean that unsettled weather is on its way.

Cirrostratus. Cirrostratus clouds are high-level clouds that cover the sky. The cloud layer is thin and transparent. You can see the Sun or the Moon through cirrostratus clouds.

Investigation 7: Water in the Air

US Map with Western Cities

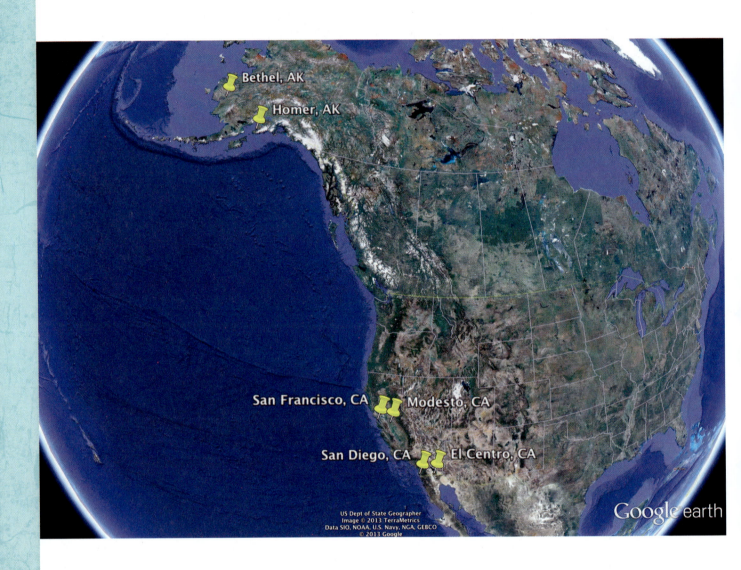

Alaska Climate over 30 Years

Homer, Alaska

Latitude: 59.6° N
Longitude: 151.5° W
Elevation: 10 m
Distance from ocean: 0 km
Average annual precipitation: 64.6 cm

Average temperature in °C

Jan	Feb	Mar	Apr	May	Jun
-5.0	-4.2	-1.7	2.2	6.7	10.0

Jul	Aug	Sep	Oct	Nov	Dec
12.2	12.2	8.9	3.1	-1.4	-3.3

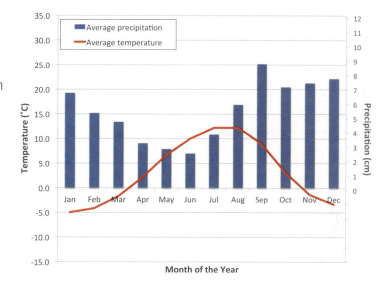

Bethel, Alaska

Latitude: 60.8° N
Longitude: 161.8° W
Elevation: 10 m
Distance from ocean: 96 km
Average annual precipitation: 38.8 cm

Average temperature in °C

Jan	Feb	Mar	Apr	May	Jun
-14.2	-13.6	-9.7	-3.6	5.0	10.6

Jul	Aug	Sep	Oct	Nov	Dec
13.3	11.9	7.5	-1.1	-8.1	-12.5

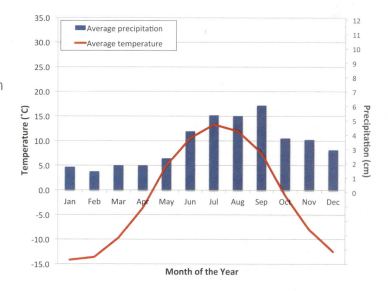

Investigation 8: The Water Planet

Southern California Climate over 30 Years

San Diego, California

Latitude: 32.7° N
Longitude: 117.2° W
Elevation: 9 m
Distance from ocean: 0 km
Average annual precipitation: 30.4 cm

Average temperature in °C

Jan	Feb	Mar	Apr	May	Jun
13.1	13.3	13.9	15.3	16.9	19.2

Jul	Aug	Sep	Oct	Nov	Dec
21.7	22.2	21.7	18.9	15.6	13.1

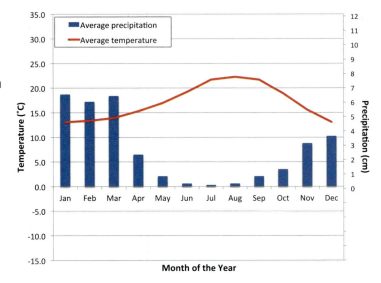

El Centro, California

Latitude: 32.8° N
Longitude: 115.6° W
Elevation: -9 m
Distance from ocean: 144 km
Average annual precipitation: 7.5 cm

Average temperature in °C

Jan	Feb	Mar	Apr	May	Jun
13.1	15.3	17.8	20.8	25.3	29.7

Jul	Aug	Sep	Oct	Nov	Dec
33.1	33.1	30.0	23.9	16.9	12.8

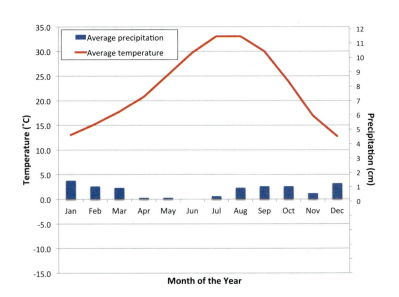

Northern California Climate over 30 Years

San Francisco, California

Latitude: 37.8° N
Longitude: 122.4° W
Elevation: 9 m
Distance from ocean: 0 km
Average annual precipitation: 56.5 cm

Average temperature in °C

Jan	Feb	Mar	Apr	May	Jun
11.1	12.8	13.1	14.2	14.4	15.8

Jul	Aug	Sep	Oct	Nov	Dec
16.1	16.9	17.5	16.9	14.2	11.7

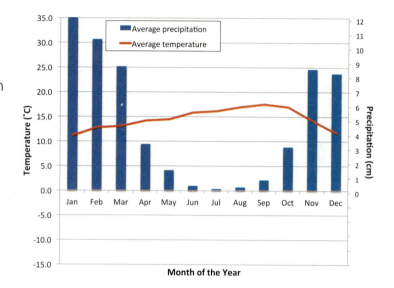

Modesto, California

Latitude: 37.6° N
Longitude: 121.0° W
Elevation: 24 m
Distance from ocean: 129 km
Average annual precipitation: 33.3 cm

Average temperature in °C

Jan	Feb	Mar	Apr	May	Jun
8.3	11.7	13.6	16.4	20.0	23.3

Jul	Aug	Sep	Oct	Nov	Dec
25.6	24.7	22.8	18.3	12.2	8.1

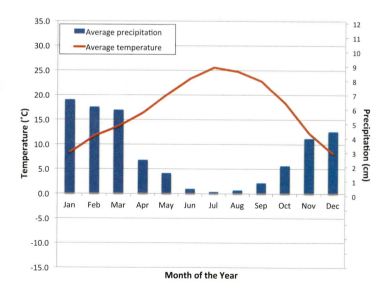

Climate Graph A

Climate Data for Denver, Colorado

1901–1930
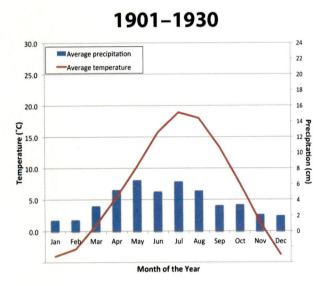

1981–2010
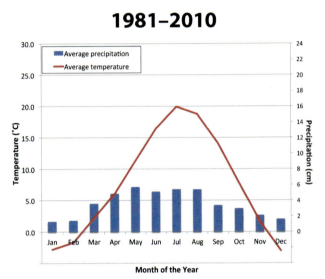

Climate Data for Miami, Florida

1901–1930
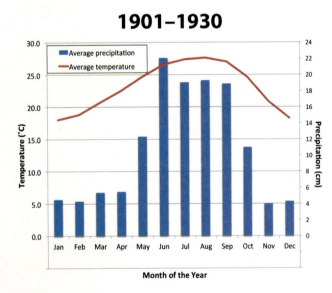

1981–2010
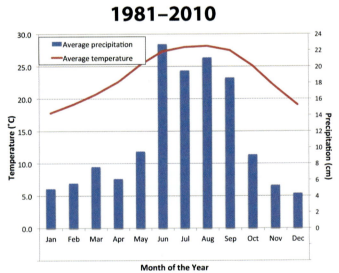

Climate Graph B

Climate Data for Los Angeles, California

1901–1930
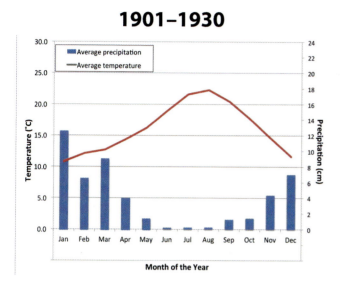

1981–2010
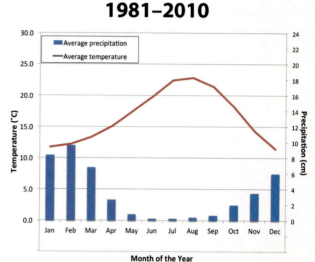

Climate Data for New York, New York

1901–1930
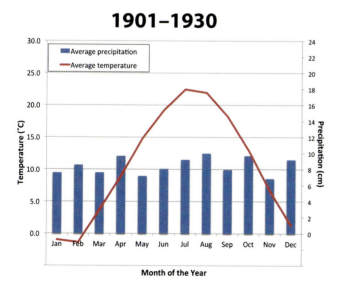

1981–2010
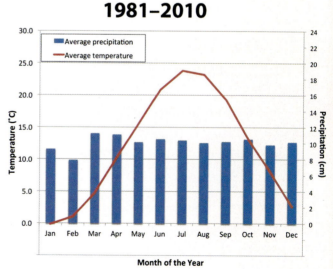

Investigation 9: Climate over Time **131**

Fronts

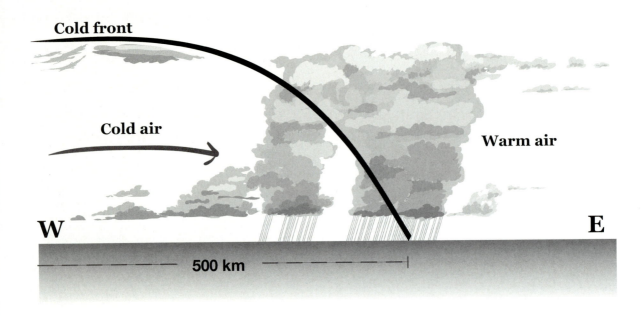

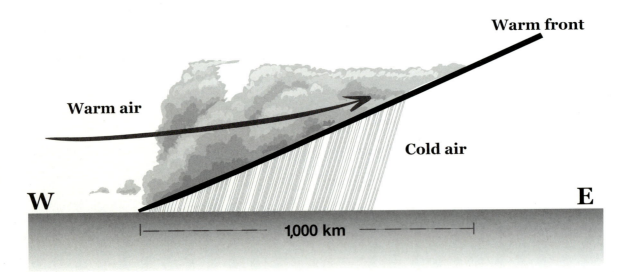

Weather and Fronts

Cold Front

Weather observation	Before front passes	While front passes	After front passes
Winds	South to southwest	Gusty, shifting	West to northwest
Temperature	Warm	Sudden drop	Drops steadily
Pressure	Falls steadily	Small drop, then sharp rise	Rises steadily
Clouds	Cloud cover increases; cirrus, cirrostratus, cumulonimbus	Cumulonimbus	Cumulus
Precipitation	Short period of showers	Heavy rain, sometimes with thunderstorms, including hail	Showers followed by clearing
Visibility	Fair to poor, hazy	Poor, then improving	Good, unless there are showers
Dew point	High and steady	Drops sharply	Continues to lower

Warm Front

Weather observation	Before front passes	While front passes	After front passes
Winds	South to southeast	Variable	South to southwest
Temperature	Cool/cold, slowly warming	Steadily rising	Warming, then steady
Pressure	Usually falling	Leveling off	Rises slightly, then falls
Clouds	Clouds usually in this order: cirrus, cumulostratus, altostratus, nimbostratus, stratus, then fog; cumulonimbus also possible in summer	Stratus and variations of stratus	Clearing followed by stratocumulus; cumulonimbus also possible in summer
Precipitation	Light to moderate rain or drizzle	Drizzle or none	Usually none, but sometimes light rain or showers
Visibility	Poor	Poor, but improving	Hazy or fair
Dew point	Steady rise	Steady	Rising, then steady

Investigation 10: Meteorology

Science Practices

1. **Asking questions.** Scientists ask questions to guide their investigations. This helps them learn more about how the world works.
2. **Developing and using models.** Scientists develop models to represent how things work and to test their explanations.
3. **Planning and carrying out investigations.** Scientists plan and conduct investigations in the field and in laboratories. Their goal is to collect data that test their explanations.
4. **Analyzing and interpreting data.** Patterns and trends in data are not always obvious. Scientists make tables and graphs. They use statistical analysis to look for patterns.
5. **Using mathematics and computational thinking.** Scientists measure physical properties. They use computation and math to analyze data. They use mathematics to construct simulations, solve equations, and represent different variables.
6. **Constructing explanations.** Scientists construct explanations based on observations and data. An explanation becomes an accepted theory when there are many pieces of evidence to support it.
7. **Engaging in argument from evidence.** Scientists use argumentation to listen to, compare, and evaluate all possible explanations. Then they decide which best explains natural phenomena.
8. **Obtaining, evaluating, and communicating information.** Scientists must be able to communicate clearly. They must evaluate others' ideas. They must convince others to agree with their theories.

Are you a scientist?

Engineering Practices

1. **Defining problems.** Engineers ask questions to make sure they understand problems they are trying to solve. They need to understand the constraints that are placed on their designs.

2. **Developing and using models.** Engineers develop and use models to represent systems they are designing. Then they test their models before building the actual object or structure.

3. **Planning and carrying out investigations.** Engineers plan and conduct investigations. They need to make sure that their designed systems are durable, effective, and efficient.

4. **Analyzing and interpreting data.** Engineers collect and analyze data when they test their designs. They compare different solutions. They use the data to make sure that they match the given criteria and constraints.

5. **Using mathematics and computational thinking.** Engineers measure physical properties. They use computation and math to analyze data. They use mathematics to construct simulations, solve equations, and represent different variables.

6. **Designing solutions.** Engineers find solutions. They propose solutions based on desired function, cost, safety, how good it looks, and meeting legal requirements.

7. **Engaging in argument from evidence.** Engineers use argumentation to listen to, compare, and evaluate all possible ideas and methods to solve a problem.

8. **Obtaining, evaluating, and communicating information.** Engineers must be able to communicate clearly. They must evaluate other's ideas. They must convince others of the merits of their designs.

Are you an engineer?

Engineering Design Process

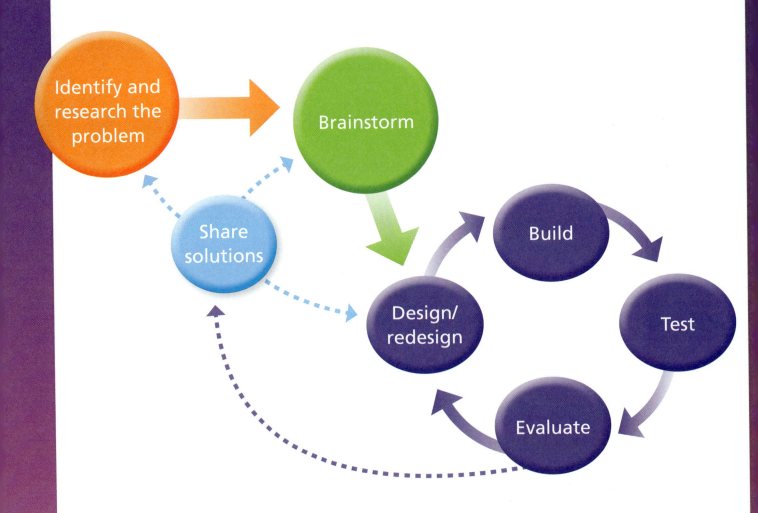

Science Safety Rules

1. Always follow the safety procedures outlined by your teacher. Follow directions, and ask questions if you're unsure of what to do.

2. Never put any material in your mouth. Do not taste any material or chemical unless your teacher specifically tells you to do so.

3. Do not smell any unknown material. If your teacher asks you to smell a material, wave a hand over it to bring the scent toward your nose.

4. Avoid touching your face, mouth, ears, eyes, or nose while working with chemicals, plants, or animals. Tell your teacher if you have any allergies.

5. Always wash your hands with soap and warm water immediately after using chemicals (including common chemicals, such as salt and dyes) and handling natural materials or organisms.

6. Do not mix unknown chemicals just to see what might happen.

7. Always wear safety goggles when working with liquids, chemicals, and sharp or pointed tools. Tell your teacher if you wear contact lenses.

8. Clean up spills immediately. Report all spills, accidents, and injuries to your teacher.

9. Treat animals with respect, caution, and consideration.

10. Never use the mirror of a microscope to reflect direct sunlight. The bright light can cause permanent eye damage.

Glossary

absorb to soak up or take in

air the mixture of gases surrounding Earth

air pressure the force exerted on a surface by the mass of the air above it; also called atmospheric pressure

atmosphere the layer of gases surrounding Earth. Its layers include the troposphere, stratosphere, mesosphere, thermosphere, and exosphere.

atmospheric pressure the force exerted on a surface by the mass of the air above it; also called air pressure

axis an imaginary axle that a planet spins on

barometer a weather tool that measures air pressure

bimetallic strip a narrow band made of two metals stuck together

blizzard a severe storm with low temperatures, strong winds, and extreme snow

boundary current a large-scale ocean current along a coastline

carbon dioxide (CO_2) a greenhouse gas in Earth's atmosphere. Carbon dioxide is created by natural and human-made processes.

cold front a zone where a faster-moving cold air mass collides with a warm air mass, resulting in brief, intense precipitation

conduction energy transfer by contact between particles

constraint a restriction or limitation

convection energy transfer during which hot fluid rises and cold fluid sinks, creating a cycle

convection cell a mass of fluid flowing in a cycle in one place

Coriolis effect the cause of winds near the equator in the Northern and Southern Hemispheres to curve west

criterion (plural: **criteria**) requirement

cumulonimbus clouds that are piled up from low to high levels and bring rain

density the amount of mass in a sample of matter compared to the volume of that sample

doldrums the calm, windless region near the equator

drought a less-than-normal amount of rain or snow over a long period of time

dust devil a small rotating wind column that becomes visible when it collects dust and debris

dust storm severe weather in which strong winds carry large quantities of dust over an area

El Niño a flow of unusually warm water in the eastern Pacific Ocean. It causes many changes in weather in other places.

equinox a day of the year when the Sun's rays shine straight down on the equator and the length of day and night are equal everywhere

exosphere the layer of the atmosphere above the thermosphere. The exosphere marks the transition from the atmosphere to space.

flash flood a short, rapid, unexpected flow of water and debris

flood a large amount of water spilling onto land that is usually dry

fluid a substance that can flow, such as a gas or liquid

global warming a worldwide warming trend on Earth that affects global weather

greenhouse gas a gas that absorbs and radiates energy in the atmosphere, trapping thermal energy

groundwater water stored below Earth's surface

Gulf Stream a warm boundary current in the North Atlantic Ocean

gyre a large system of rotating ocean currents

Hadley cell a huge convection cell that covers much of Earth at the equator

hail precipitation in the form of small balls of ice

horse latitudes the windless areas around 30° north and south of the equator

hurricane a cyclone or severe rotating tropical storm that produces high winds and severe rain in the Northern Hemisphere east of the International Date Line

infrared radiant energy that is beyond the red end of the visible spectrum

insulation material that can reduce energy transfers

jet stream a narrow band of fast-moving high-altitude wind flowing west to east that affects weather conditions on Earth

kinetic energy energy of motion

land breeze a wind that blows from land to sea

latitude the distance north or south from the equator measured in degrees; a factor that affects local weather and climate

lightning a bright flash of light caused by an electric discharge between two clouds or from a cloud to Earth

mass the amount of matter in an object or sample

matter anything that has mass and takes up space

mesosphere the layer of the atmosphere above the stratosphere

methane a variable gas in the atmosphere; a greenhouse gas

microburst a short, very intense downdraft of air

model a representation of thinking to help others understand your thinking

nitrogen a colorless, odorless gas that makes up about 78 percent of Earth's atmosphere

nonrenewable a material that cannot be replaced once used up

North Star the reference star pointed to by the North Pole

ocean current a global water pattern affected by winds, differences in water density, tides, and landmasses

orbit the path and time it takes one object to travel around another object

oxygen a gas that makes up about 21 percent of Earth's atmosphere

ozone a form of oxygen that forms a thin layer in the stratosphere

permanent gas a gas in the atmosphere for which the relative amount stays constant. Oxygen and nitrogen are permanent gases.

photosynthesis a process used by plants and algae to make sugar (food) out of light, carbon dioxide, and water

planet an object that orbits a star and is massive enough for its own gravity to force it into a spherical shape

prevailing wind a predictable wind produced by the combination of high- and low-pressure areas and the Coriolis effect.

radiant energy energy that travels through air and space

radiosonde an instrument sent into Earth's atmosphere to measure temperature, air pressure, humidity, and wind

renewable able to be replaced or restored by nature

revolution one complete orbit around something

rip current a strong local ocean current that moves directly away from shore

rotation spinning on an axis

salinity the amount of dissolved salt in water

sea breeze a wind that blows from sea onto land

season a period of the year identified by changes in hours of daylight and weather

severe weather out-of-the-ordinary and extreme weather conditions

solar energy radiant energy from the Sun

solstice a day of the year when the North Pole is angled farthest toward or away from the Sun

star a large, hot ball of gas radiating huge amounts of energy

step leader a downward path of electric charge from a cloud to Earth, producing lightning

straight-line wind a strong wind that has no rotation

stratosphere the layer of the atmosphere above the troposphere. The ozone layer is in the stratosphere.

thermal energy radiant energy that heats

thermometer an instrument that measures thermal energy as temperature

thermosphere the layer of the atmosphere above the mesosphere

thunder a loud, explosive sound created by lightning

thunderstorm severe weather that results from cold air flowing under a warm, humid air mass over the land, resulting in lightning, thunder, heavy precipitation, and possible tornadoes

tornado a rapidly rotating destructive column of air that extends from a thunderstorm to the ground. Wind speeds can exceed 400 kilometers per hour in a tornado.

trade wind the predictable moderate wind between 5° and 25° north or south latitude

troposphere the layer of the atmosphere that begins at Earth's surface and extends upward for an average of 15 kilometers. Weather happens in the troposphere.

typhoon a cyclone or severe tropical storm that produces high winds in the Pacific north of the equator and west of the International Date Line

updraft a forceful vertical air current

vacuum space containing no particles of air or anything else

variable gas an atmospheric gas whose amount changes based on the environment. Carbon dioxide, methane, water, and ozone are variable gases.

vortex a tornado-like swirling column of wind or water

warm front a zone where a faster-moving warm air mass collides with a cold air mass, resulting in prolonged, light precipitation

water vapor the gaseous state of water; a variable gas that is found in Earth's atmosphere

weather factor a variable property of weather, such as temperature, humidity, air pressure, or wind

wildfire a fire occurring in nature that can be driven by winds

When meteorologists describe weather, they consider several weather factors, or properties, such as temperature, wind, and precipitation, like the rain shown here.

Index

A
Absorb, 22, 138
Air, 10, 18–23, 26, 27, 46, 52, 61–68, 76–84, 85–86, 138
Air pressure, 31, 32–40, 83–84, 85, 138
Atmosphere, 8, 26–31, 69–75, 138
Atmospheric pressure, 12, 31–40, 138
Axis, 54, 138

B
Barometer, 37, 38, 39, 138
Bimetallic strip, 61, 138
Blizzard, 14–15, 138
Boundary current, 99–100, 138

C
Carbon dioxide, 18–22, 72, 75, 106, 109–110, 138
Celsius, Anders, 60
Celsius scale, 60
Climate change, 75, 105–110
Cold front, 133, 138
Condensation, 21
Conduction, 64, 68, 70–71, 73, 78, 138
Constraint, 65, 138
Convection, 51–52, 64, 74, 76–77, 78–79, 96–97, 138
Convection cell, 52, 74, 76–79, 81, 138
Coriolis effect, 80, 98, 138
Criteria, 65, 138
Cumulonimbus, 6, 138
Current, 96–100, 103, 105–106

D
Day/Daylight, 53–58, 60
De' Medici, Ferdinando II, 60
Density, 36, 41–46, 47–50, 63, 96–98, 138
Doldrums, 81, 138
Drought, 5, 16, 94, 108, 138
Dust devil, 16–17, 138
Dust storm, 16–17, 138

E
Earth, 24–31, 54–58, 76–84, 91–95, 107–108, 109–110
El Niño, 103–104, 138
Energy transfer, 46, 52, 64, 70–71, 73
Engineering design process, 136
Engineering practices, 135
Equinox, 55, 57, 58, 138
Evaporate, 21, 94
Evidence, 89
Exosphere, 30, 139

F
Fahrenheit, Daniel Gabriel, 60
Fahrenheit scale, 60
Flash flood, 4, 139
Flood, 3, 4, 12, 13, 139
Fluid, 52, 139
Fossil fuel, 106, 110

G
Galilei, Galileo, 63
Gas, 18–23, 34–35, 42, 45, 46, 72, 75, 106, 107, 109–110
Global warming, 94, 106–108, 139
Greenhouse gas, 21, 23, 72–73, 75, 139
Groundwater, 5, 139
Gulf stream, 100, 139
Gyre, 96–102, 139

H
Hadley, George, 79
Hadley cell, 79–81, 139
Hail, 6, 7, 139
Heat, 51–52, 64, 69–75, 80, 82, 83
Helium, 20, 41–45
Horse latitude, 81, 139
Humidity, 27, 85, 86
Hurricane, 3, 12–13, 16, 88, 108, 139
Hydrogen, 18, 20, 23, 85

I
Infrared, 71, 72, 73, 139
Insulation, 64–68, 139

J
Jet stream, 28, 81, 139

K
Kinetic energy, 34, 46, 52, 68, 70, 71, 139

L
Land breeze, 84, 139
Latitude, 58, 139
Lightning, 8–9, 139

M
Mass, 20, 28, 32, 34, 42–45, 47–50, 52, 139
Matter, 37, 42–46, 55, 139
Mesosphere, 29, 30, 139
Meteorologist, 8, 27, 36–37, 38, 40
Methane, 23, 139
Microburst, 16, 139
Model, 65, 77, 79, 139

N
Nitrogen, 19, 20, 34, 42, 73, 139
Nonrenewable, 66, 139
North Star, 55, 56, 139

O
Ocean current, 96–102, 103, 105, 140
Orbit, 54–55, 56, 90, 140
Oxygen, 19, 20, 22, 28, 36, 42, 72, 73, 140
Ozone, 19, 20, 22, 28, 140

P
Permanent gas, 20, 140
Photosynthesis, 22, 140
Planet, 17, 18–19, 52, 74, 78–79, 91–95, 140
Prevailing wind, 76, 80, 81, 140

R
Radiant energy, 71, 140
Radiation, 22, 28, 29, 64, 71–73
Radiosonde, 85, 140
Renewable, 66, 140
Revolution, 54, 140
Rip current, 99, 140
Rotation, 16, 54–55, 98, 140

S
Safety rules, 137
Salinity, 97, 140
Science practices, 134
Sea breeze, 83, 140
Season, 27, 53–58, 140
Severe weather, 3–23, 140
Solar energy, 56–57, 73, 78, 140
Solstice, 55–58, 140
Star, 18, 140
Step leader, 8, 140
Straight-line wind, 16, 140
Stratosphere, 22, 28, 29, 30, 140
Sun, 18, 22, 24, 26–29, 53–58, 71–72, 98

T
Temperature, 24–30, 59–63, 73, 82–84, 85, 97, 103, 105, 108–109
Thermal energy, 21, 72–74, 106, 140
Thermometer, 59–63, 140
Thermosphere, 29–30, 140
Thunder, 8–9, 140
Thunderstorm, 3, 7–11, 16, 141
Tornado, 3, 10–11, 15, 16, 88–89, 141
Torricelli, Evangelista, 37
Trade wind, 81, 102, 141
Troposphere, 27, 25, 27, 33, 141
Typhoon, 3, 12, 141

U
Updrafts, 6, 141

V
Vacuum, 37, 141
Variable gas, 21, 141
Volume, 20, 21, 42–46, 47–50, 59
Vortex, 88, 141

W
Warm front, 133, 141
Water, 4–5, 21–23, 41–45, 51–52, 60, 63, 72–74, 87–98, 103–104
Waterspout, 88–89
Water vapor, 19–21, 72–74, 78, 92, 141
Weather, 3–17, 26, 27, 40, 76, 85–86, 94, 96, 103–104
Weather balloon, 27, 39, 85–86
Weather factor, 27, 141
Wildfire, 17, 141
Wind, 7, 10–17, 27, 28, 76–85, 88, 89, 96, 103
Windstorm, 16–17